DATE			

METEORITES

METEORITES

****

Their Record of Early

Solar-System History

****

JOHN T. WASSON
University of California, Los Angeles

W. H. Freeman and Company
New York

Library of Congress Cataloguing in Publication Data

Wasson, John T.
  Meteorites: their record of early solar-system history.

  Bibliography: p.
  Includes index.
  1. Meteorites. 2. Solar-system —
History. I. Title.
QB755.W374 1985   523.5′1   85-1484
ISBN 0-7167-1700-X

Printed in the United States of America.

1 2 3 4 5 6 7 8 9 0 MP 3 2 1 0 8 9 8 7 6 5

Contents

Preface

This book is written as a text for a general course covering meteoritic evidence regarding the formation and early evolution of the solar system. It is less detailed and less technical than my 1974 book, *Meteorites: Their Classification and Properties,* and covers additional topics including the formation of the different kinds of meteorites and the compositional evidence linking meteorite groups with individual planets. Meteorites are the chief source of information about the earliest period of solar-system history. My goal is to make this information available in a relatively inexpensive form to those interested in learning more about this exciting interdisciplinary research field.

Any author of a text aimed at a broad scientific audience must decide how much scientific jargon to include. Although I have made a conscious effort to avoid unnecessary jargon, I have left in many words for which the substitution of a more-readily understood term would result in a significant degradation of meaning or precision. For example, I have used mineral names where it seemed necessary; in some of these cases I have also given the chemical formula. I have used the symbols of the chemical elements throughout the text; the reader unfamiliar with these will find them deciphered in Appendix D. I have used a set of units closely related to the International System (SI) including the SI prefixes (such as G for 10^9); these units and prefixes are listed in Appendix B. A number of words that are common in the meteoritic or astronomical literature but are either uncommon or have altered meanings in other fields are indicated by **boldface** type where defined in the text and are similarly marked in the index.

The reader should note that I use some notations that are not yet common in the scientific literature. For consistency, all concentration units are SI: mg/g, μg/g, ng/g, and so on. These units have the distinct advantage of always differing by the same factor of 10^3; for this reason, wt% is not used.

A problem any author of a scientific text faces is how to handle citations. I have chosen to follow the practice common in textbooks of limiting citations to those materials appropriate for supplementary reading by a general reader and to those papers whose data or diagrams are cited directly. This makes it difficult for a reader to trace the sources of ideas and to find those occasional papers that give interpretations different from those mentioned here, but in my opinion the advantage of a more-readable text outweighs this disadvantage.

I am indebted to many persons for assistance. Sue Hamilton, Vicki Doyle Jones, Kaye Lee, Anne Young, and especially Grace No patiently typed and drafted the many versions of the text and figures. Numerous persons and institutions allowed their photos or drawings to be reproduced. Several of my research colleagues (especially Don Brownlee, Jeff Grossman, Greg Kallemeyn, Alan Rubin, Dave Shirley, Paul Warren, and John Willis) have suggested improvements in the text or helped in the preparation of diagrams. To all of these and the many others who have shared their knowledge of meteorites with me, I am most grateful. My greatest debt is to Gudrun, Gisla, and Kerstin Wasson for their warm support and for sparing me so many evening and weekend hours.

John T. Wasson
Los Angeles
October 1984

METEORITES

CHAPTER I

Meteorite Recovery, Fall Phenomena, Craters, and Orbits

Survey of the Solar System

The Moon and hundreds of artificial satellites are in orbit about the Earth; the Earth, the other planets, the asteroids, and the comets are in orbit about the Sun; and the Sun is in orbit about the center of the Milky Way galaxy. The basic equations describing the motions of these bodies are discussed in Appendix H.

Figure I-1 shows the orbits of the planets projected onto the ecliptic plane, the plane of the Earth's orbit; an inset shows the orbits of the inner planets at a larger scale. The planets are commonly divided into the inner, or terrestrial, planets (Mercury, Venus, Earth, and Mars) and the giant, or jovian, planets (Jupiter, Saturn, Uranus, and Neptune). Pluto does not seem to be a true planet, but rather a member of the set of smaller objects that accumulated to form Uranus and Neptune. Thousands of rocky bodies having radii in the range of 1 to 500 km[1] occupy orbits between those of Mars and Jupiter; a small fraction of these **asteroids**[2] are in other orbits, some of which cross the Earth's orbit. **Comets** are icy bodies that partially evaporate in the inner solar system to produce extended atmospheres; most comets are in highly elliptical orbits having long axes much greater than that of Neptune's orbit.

Most of the mass of the solar system is in the Sun, and most of the mass of the planetary system is in Jupiter. It is often handy to remember the fact that the mass of the Sun is ~ 1000 times[3] that of Jupiter, the mass of Jupiter ~ 1000 times that of the Earth, and the mass of the Earth ~ 1000 times that of the Ceres, which is by far the most massive asteroid. The properties of the planets, asteroids, and comets are discussed in more detail in Chapter IX.

Meteorites and Meteoroids

The solar system includes not only the planets, asteroids, and comets, but also solid objects ranging in size from bodies with radii of several kilometers

---

[1] See Appendix B for a list of the units used in this book.

[2] Definitions of specialized words are indicated by using **boldface** type for the word being defined.

[3] The symbol $\sim$ before a number indicates an approximate value.

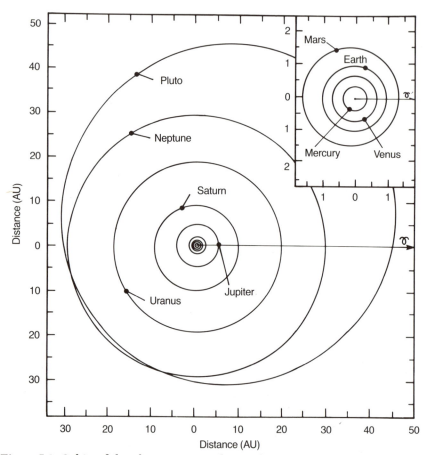

Figure I-1. Orbits of the planets projected onto the Sun–Earth (ecliptic) plane. Because the radii of the orbits of the planets increase exponentially with increasing distance from the Sun, it is not convenient to show them all on a single diagram. For this reason, the orbits of the inner planets are enlarged in the inset diagram. The orbits are nearly circular, except those of Mars and Pluto (which should no longer be classified as a planet). The arrow shows the direction from the Earth to the Sun at the time of the vernal equinox.

to submicroscopic dust. Some of these smaller objects are in **heliocentric** (Sun-centered, or planetlike) orbits that intersect that of the Earth. Such an object is called a **meteoroid** before it enters the Earth's atmosphere, a **meteor** when atmospheric friction makes it incandescent, and a **meteorite** if it is recovered. Meteorites recovered following observed falls are called **falls**; those recognized in the field that cannot be definitely associated with observed falls are called **finds**.

Meteorites having masses greater than 500 g fall at a rate of about one per 10^6 km^2 per year. Although about 150 fall on the land every year, only a few are recovered immediately following the fall. As a result, the chances are very small that an outdoors person such as a farmer will witness the fall of a subsequently recovered meteorite during a lifetime; the chances for an urban dweller are much smaller still.

As discussed in more detail in Chapter II, most meteorite falls are composed primarily of silicate minerals and are designated **stony meteorites** — or, where the meaning is clear from the context, simply "stones." Because they have a greater resistance to weathering and are more readily recognized in the field, meteorites composed largely of Fe–Ni metal are especially common among finds. These meteorites are designated **iron meteorites,** or simply "irons."

The Beginnings of Meteorite Study

Meteorite falls have been witnessed throughout human history. Iron meteorites were an important raw material to primitive peoples. In some cases, the heavenly source of this iron was recognized, as evidenced by etymological relationships.

Figure I-2 is a charming woodcut showing the fall of a meteorite in Ensisheim, Alsace, in 1492 — a year well known to school children in the

Figure I-2. Woodcut depicting the fall of the Ensisheim LL chondrite on 7 November 1492. A literal translation of the German caption (by Sebastian Brant) is "of the thunder-stone (that) fell in xcii (92) year outside of Ensisheim." This meteorite, which is preserved in the city hall of Ensisheim, Alsace, is the oldest recorded fall from which material is still available.

Americas. In the original broadsheet, the major German author Sebastian Brant provided a verse description of the fall in Latin and German. The verse suggests that the Austrian king, Maximilian, won a battle against the Burgundians because the latter were terrified by the extraterrestrial fireworks above their heads. This meteorite was clearly recognized to have fallen from the sky.

However, during the eighteenth century (the "age of enlightenment"), scientists became convinced that reports of rocks from the sky were but the fantasies of peasants. Perhaps the situation arose because these learned men had never witnessed a meteorite fall, perhaps more so because their attempts to eliminate superstition led them to a general skepticism regarding the testimony of peasants and other inhabitants of rural places. In any case, it was possible for P. Bertholon to comment regarding the 1790 fall of a meteorite at Barbotan, France, "how sad it is that the entire municipality enters folk tales upon the official record." A. Stutz in 1790, in describing the fall of an iron meteorite near Hraschina, Croatia (Yugoslavia), noted "the straightforward manner with which everything is accounted, the agreement among the witnesses who had no grounds to agree so completely regarding a lie," but then concluded that "in our times it would be unforgivable to hold such tales to be probable." It has been widely quoted that, on hearing of the description of a meteorite fall at Weston, Connecticut, on 14 December 1807, Thomas Jefferson said, "It is easier to believe that two Yankee professors would lie, than that stones would fall from heaven." However, recent searches have failed to locate the source of this reputed quotation, and there is growing doubt that Jefferson actually made the statement. A skeptical but much more balanced statement, however, is found in Jefferson's letter to Daniel Solomon on 15 February 1808: "its [the Weston stone's] descent from the atmosphere presents so much difficulty as to require careful examination," but "we certainly are not to deny whatever we cannot account for," and "the actual fact . . . is the thing to be established."

The first clear and detailed exposition of the extraterrestrial nature of meteorites is found in a small book published by E. F. F. Chladni in 1794. Chladni not only showed that there was good factual support for the conclusion that bright fireballs often drop meteorites, but he also gave well-reasoned arguments for believing that certain large masses of iron are prehistoric meteorites. Soon after the distribution of his book, Chladni's ideas were supported by the publication of several reports of meteorite falls and by the discovery of the first two asteroids in 1801 and 1802. The fraction of the scientific community that held meteorites to be extraterrestrial, although minor, was clearly increasing.

It is commonly stated that the turning point occurred in 1803, when J. B. Biot, a young member of the Académie Française, investigated and strongly endorsed the correctness of reports that a shower of meteoritic stones had fallen from the sky at L'Aigle, France. In fact, the tide had started to turn a few years earlier. Although a few ingenious persons still hypothesized that meteorites are atmospheric condensations, these ideas faded during the following decades.

Fall Phenomena

The fall of a meteorite is typically a spectacular event. The meteoroid enters the atmosphere with a velocity of about 20 km · s^{-1}. At an altitude of about 100 km, atmospheric density is great enough to produce a significant amount of frictional drag. The kinetic energy lost in the deceleration of the meteoroid is converted primarily to heat and light. The brightness of the fireball is commonly great enough to be seen in daylight. Nighttime fireballs that reach the earth as meteorites are in most cases brighter than the full Moon.

Heating by the impacting air molecules melts the surface of the meteoroid. Selective erosion produces attractive hills and valleys called **regmaglypts,** particularly on iron meteorites. If a meteoroid maintains a fixed orientation during atmospheric passage, the regmaglypts on the leading surface become much deeper than those on the trailing surface. The iron from Cabin Creek, Arkansas, is the classic example of an oriented meteorite (Figure I-3). By the time the meteoroid reaches terminal velocity (a nearly constant velocity of ~0.3 km · s^{-1} at which the deceleration by drag is approximately balanced by the acceleration by gravity), frictional heating is no longer sufficient to produce melting, and the final melt has congealed to form a fusion crust. Because heat conduction is a slow process, the interiors of meteoroids remain cold even when their surfaces are molten. Significant thermal effects usually extend inward a few centimeters in the efficiently conducting irons, but only several millimeters in the poorly conducting stones. The heat-altered zone of an iron-meteorite fall is readily recognized on a polished surface (Figure I-4).

The meteoroid is subjected to severe stresses during the deceleration to subsonic speeds from velocities many times supersonic. If it consists of **friable** (easily crumbled) material, it will disintegrate into individual grains and not penetrate as a rock-sized body to the Earth's surface. Even if it is tough, the stresses generally break up stones larger than several kilograms or irons larger than several tens of kilograms, resulting in a meteorite shower. The fracturing commonly occurs along preexisting cracks produced by impacts with other space debris prior to capture by the Earth. The shower of meteorites results in a **strewn field,** a set of meteorite fragments scattered over a certain geographic region that is typically elliptical in

Figure I-3. The two sides of the Cabin Creek, Arkansas, IIIAB iron are strikingly different, apparently because the meteoroid remained in a fixed orientation during atmospheric passage, with the face shown above leading and that on page 7 trailing. The total height is 44 cm. A meteoritic mass that was worshiped

in Ephesus is described in the biblical Acts of the Apostles. It seems likely that the interpretation of this meteorite as an Earth Goddess was prompted by a benippled appearance similar to Cabin Creek's dimpled leading side. (Photo from G. Kurat, Naturhistorisches Museum, Vienna)

Figure I-4. A slice through the Avce, Yugoslavia, IIAB iron, an observed fall. The polished and etched surface of this hexahedrite is crossed by three sets of Neumann lines (shock-produced crystallographic twins) having different orientations. Frictional heating during atmospheric deceleration has produced a heat-altered zone free of Neumann lines around the edge of the section. (Photo from V. F. Buchwald, *Iron Meteorites*, University of California Press, 1975)

shape (Figure I-5). Because of their lower surface-to-mass ratios, larger stones travel farther than smaller stones along the "great circle" projection of the trajectory along the Earth's surface. The deceleration of the meteoroid also generates sonic booms. These thunderlike noises are heard chiefly by persons directly below the trajectory.

If they remain intact, the largest meteoroids are never fully decelerated to terminal velocity in the air; they retain some fraction of their preatmospheric velocity v_∞ at impact with the solid Earth. Figure I-6 shows the approximate theoretical relationship between the relative velocity v/v_∞ and a parameter K defined as

$$K = \frac{0.7D}{r\rho \sin \theta}$$

where D is the drag coefficient (which can take on values between 0 and 1), r is the radius in cm, ρ is the density in g · cm^{-3}, and θ is the atmospheric entry angle between the meteoroid's trajectory and the Earth's surface. For

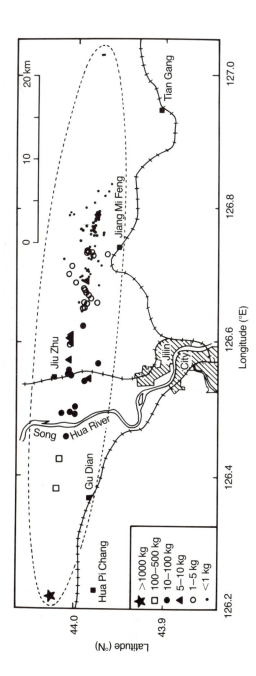

Figure I.5. Distribution of fragments from the shower of H chondrites near Jilin City, Jilin Province, China, on 8 March 1976. The meteoroid's direction of flight was toward the west. After disintegration in the atmosphere, the larger the fragment, the farther it traveled before striking the Earth. The resulting dispersion ellipse is typical of meteorite showers.

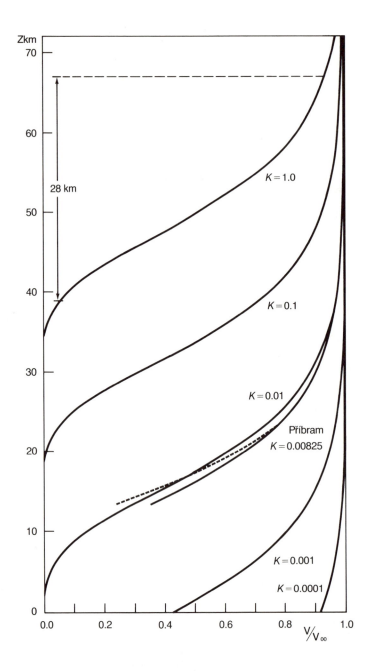

Figure I-6. Theoretical curves predicting the ratio of the velocity v of a meteoroid relative to the velocity v_∞ that it had immediately before striking the atmosphere, as a function of the altitude z and a parameter K that is inversely proportional to both radius and density. The dashed curve shows that the photographically determined path of the Příbram H5 chondrite is approximately matched by the appropriate $K = 0.00825$ curve. (From V. F. Buchwald, *Iron Meteorites*, University of California Press, 1975)

example, a 4-ton meteorite with a drag coefficient of 0.5 entering vertically ($\theta = 90°$) would follow the $K = 0.0001$ curve and strike the Earth's surface with about 92 percent of its preatmospheric velocity. When a meteoroid strikes the Earth, it makes a hole. The typical small meteorite falling onto soil at a subsonic terminal velocity produces a pit having a volume comparable to or somewhat larger than that of the meteorite. The whole or fragmented meteorite can be recovered from such a pit.

Craters and Other Phenomena Produced by Large Meteoroids

A meteoroid that impacts the surface with nearly its preatmospheric velocity produces an explosion; during such a **hypervelocity** impact, the solid matter splashes like a fluid, and a crater much larger than the meteoroid is excavated. At an impact velocity of 20 km · s^{-1}, the radius of the crater is ~10 times and the volume nearly 1000 times that of the meteoroid. The projectile is almost entirely vaporized on impact.

The best-preserved hypervelocity crater is Meteor Crater in northern Arizona, about 30 km west of Winslow (Figure I-7). Its diameter is approximately 1200 m. A few tens of tons of iron meteorite fragments have been recovered from the projectile, whose mass was roughly 4×10^{12} g prior to

Figure I-7. An IAB iron meteorite with a radius of about 50 m created Meteor Crater in the Coconino Sandstone of Northern Arizona ~25,000 to ~50,000 years ago. The squarish outline results from preferred fracturing along roughly perpendicular zones of weakness; most craters are more circular. The mean diameter is ~1200 m. The iron meteorites associated with the crater are named after the sinuous Canyon Diablo, which crosses the upper part of the photo. (Photo by D. J. Roddy, U.S. Geological Survey)

Table I-1
Meteorites found near hypervelocity explosion craters

Meteorite	Group	Crater diameter (m)	Estimated age (kyr)
Boxhole, Northern Terr., Australia	IIIAB	160	5
Canyon Diablo, Arizona, U.S.A.	IAB	1200	20–50
Henbury, Northern Terr., Australia	IIIAB	180	<5
Kaalijarv, Estonia	IAB	100	5
Monturaqui, Atacama, Chile	IAB	370	>100
Odessa, Texas, U.S.A.	IAB	160	50
Wabar, Saudi Arabia	IIIAB	100	<5
Wolf Creek, West. Aust., Australia	IIIAB	840	>100

Source: J. T. Wasson, *Meteorites,* Springer Verlag, 1974.

impact. Recognizable meteoritic debris has been found associated with eight hypervelocity craters. In each case, the projectile belonged to one of the two largest groups of iron meteorites (Table I-1). The absence of craters produced by stones partly reflects the fact that stones weather faster than irons, but it appears to result mainly from the extensive fragmentation of large stony projectiles in the atmosphere prior to impact.

Many impact craters not associated with meteorites have been recognized on the basis of their shapes, shock-produced features, and absence of evidence for a volcanic origin. Although surface erosion tends to destroy most morphological features within a megayear or less, some very old craters are known (particularly in Canada) that were protected by sediments until recent times, then excavated by selective erosion of the sediments.

Radiometric dating of rocks returned by the Apollo missions to the Moon has revealed that the cratering rate was ~ 100 times greater ~ 4 Gyr ago than it is at present. The Earth was bombarded at that time by the same population of projectiles, so cratering may have played an important role on Earth in excavating and redistributing surficial materials, or in producing cracks through which lava could stream to the surface. A major impact is credited with triggering the magmatic activity responsible for the world's largest deposit of nickel ore near Sudbury, Ontario, Canada.

At 7:17 A.M. on 30 June 1908, an explosion occurred in a remote area of Siberia near the headwaters of the Podkamennaya (Stony) Tunguska River, at about 61°N and 102°E. The air-pressure wave was recorded on seismographs at numerous locations. The region was inhabited mainly by nomadic tribes; because of the remoteness and the unsettled political climate in Russia, no scientific studies of the immediate area were made for nearly two decades. Investigations conducted by Soviet investigators since 1927 have revealed that the explosion flattened trees over a roughly elliptical region of dimensions 40 × 50 km, and that the exteriors of trees were scorched within a smaller region 15 × 25 km (Figure I-8). Microbarograph data recorded all around the globe indicate that the energy released during the explosion was $\sim 4 \times 10^{16}$ J, equivalent to that released by exploding 10 Mt of TNT. This amount of energy is ~ 500 times greater than that of an early atomic bomb and is comparable to that of a medium-sized thermonuclear weapon.

The Tunguska explosion produced no crater, and no meteorites have been recovered. A few objects believed to be of extraterrestrial origin have been separated magnetically from local soils or peat, but it is not yet possible to confirm that these are fragments of the projectile. The absence of craters or meteorites fits together with the results of models of the forest devastation to indicate that the explosion occurred at an altitude of about 10 km above the Earth's surface.

Eyewitnesses' observations show that the Tunguska object entered the atmosphere about 660 km south-southeast of the site of the final detonation. The exact velocity cannot be inferred from these observations, but it is estimated to lie in the range of 28 to 47 km · s^{-1}.

How was it possible for the Tunguska object to deposit its entire energy in the atmosphere? Very fanciful suggestions have been made — that the object consisted of antimatter that reacted with ("annihilated") normal matter, or that it was a nuclear-powered alien spacecraft that exploded — but there is no hard evidence favoring such models. It is more likely that the object exploded in the atmosphere because it was friable and underwent total disruption during atmospheric deceleration. A large strong body reaches the Earth's surface at nearly its original velocity, but the same material dispersed as tiny grains can be stopped and its entire kinetic energy converted to thermal energy in the atmosphere. It is commonly accepted that this weak object was a fragment of a comet, and this hypothesis has recently received support from calculations showing that the object's orbit was, like the Taurid meteor streams, consistent with it being a fragment of Encke's comet. Comets are thought to consist of roughly equal parts of silicate grains and water ice, but most ice has evaporated from comets such as Encke that have made repeated passages about the Sun. It seems likely that the Tunguska object consisted mainly of an easily crushed mass of small silicate grains, and it is reasonably probable that it was cometary debris.

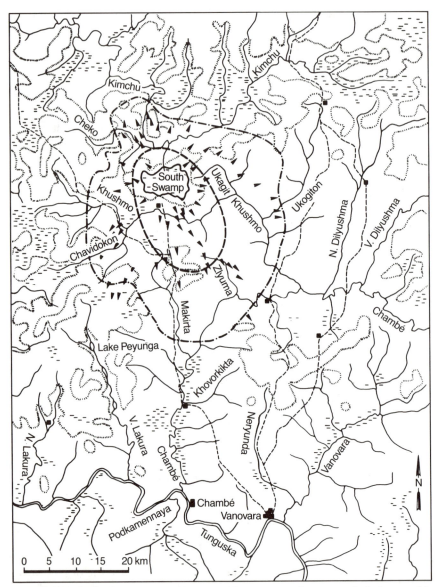

Figure I-8. The Tunguska object was traveling NNW when it exploded over this region labeled South Swamp in a heavily forested area of central Siberia. The radiation from the explosion scorched the trees inside the 15 × 25 km ellipse, and the blast wave flattened trees in the irregularly shaped region having dimensions of about 40 × 50 km. The short arrows show the directions that the trees fell. No meteorites have been recovered, but some particles extracted from the soil may be portions of the meteoroid. The object apparently consisted of weak material that was crushed and disaggregated during atmospheric passage; it probably originated in a comet.

Another remarkable meteoroid that did not reach the Earth's surface was the Rocky Mountains fireball of 10 August 1972, which made a grazing encounter with the Earth's atmosphere before returning to interplanetary space. This object was dazzlingly brilliant, even though seen on a bright early-summer afternoon. It was moving from south to north and was first observed over central Utah and last seen over central Alberta—a distance of ~ 1500 km. It was recorded by the infrared radiometer of a U.S. defense satellite, and these data combined with visual observations from the ground allowed the calculation of an accurate trajectory and orbit. The meteoroid dipped down to within 58 km of the Earth's surface over Montana; sonic booms were heard in this region. Because the trajectory was confined to the thinner part of the atmosphere, its velocity was only reduced 0.8 km $\cdot$ s^{-1} by friction—from 15.0 km $\cdot$ s^{-1} at entry to 14.2 km $\cdot$ s^{-1} at exit from the atmosphere. Estimates of its mass based on the velocity change and its luminosity are roughly congruent at a few thousand tons. Had the object hit the Earth's surface, the energy released would have been ~ 4×10^{12} J, or ~ 1 kt TNT equivalent—comparable to a very small atomic bomb. If the object had been traveling perpendicular to the Earth's surface and had survived passage through the dense part of the atmosphere without fragmentation, it would have excavated a crater with a diameter of 30 to 35 m.

Tektites are curious, small (generally less than 5 cm across), glassy objects that commonly show heat-altered zones characteristic of atmospheric passage. They appear to be melted ejecta produced by major impacts onto silica-rich sedimentary terrains. The absence of silica-rich lunar rocks effectively rules out a suggested lunar origin for these objects. The explosions apparently were so large that the expanding vapor cloud "blew off" the overlying atmosphere and launched the tektites into arching ballistic trajectories that transported them hundreds of kilometers away from the impact site. In a few cases, the parent crater is known. For example, the so-called moldavites recovered in Bohemia, Czechoslovakia, appear to be ejecta from the Ries Crater in southeastern Germany. Other possible associations are listed in Table I-2. The tektites consist almost entirely of terrestrial materials, though in some cases a bit of "meteoritic spice" is present.

The best indicators of meteoritic spice in a terrestrial sample are the siderophile elements (those concentrated in a metallic phase if one is present), especially the noble-metal subset of the siderophiles. Siderophile abundances are very low in crustal rocks because most of the siderophiles originally accreted to the Earth were extracted into the core when it formed, whereas they are present at solar-abundance levels (relative to silicon) in most of the extraterrestrial material accreted to the Earth. One noble metal, iridium (Ir), has abundances in crustal rocks lower than many other noble metals, and its concentration can be determined by the technique of neutron activation at concentrations as low as 1 pg/g (10^{-12} g/g); it

Table I-2
Tektite strewn fields and probable source craters

Tektites	Location	Age (myr)	Crater Name	Crater Location
Australites°	S. Australia			
Indochinites°	Vietnam, Cambodia, Thailand, etc.	0.71	Not known	——
Philippinites°	Philippine Isls.			
Others°†	Indonesia			
Ivory Coast tektites	Ivory Coast	1.15	Lake Bosumtwi	6.5°N 1.4°W
Moldavites	Czechoslovakia	14.0	Nordlinger Ries	48.9°N 10.6°E
Bediasites	Texas, Georgia	34.0	Not known	——
Microtektites‡	Many oceanic locations	Several vintages	——	——

° The groups listed in brackets together comprise the Australasian strewn field. Available evidence indicates that they are cogenetic.

† Rizalites, billitonites, javanites, etc.

‡ Microtektites are associated with the Australasian, Ivory Coast, and North American (bediasite) tektite-producing events and are occasionally found at other levels ("ages") in the sedimentary record.

is therefore the most-commonly used tracer of small amounts of extraterrestrial matter. Much of the iridium in sediments deposited far from land in the central part of the Pacific Ocean is of extraterrestrial origin.

A spectacular discovery of a few years ago is a thin clay layer with exceptionally great iridium concentrations that was deposited all around the world at the boundary between the Cretaceous geologic period (the age of dinosaurs) and the Tertiary geologic period (the age of mammals). The most-plausible explanation of this observation is that a large (5 to 10 km radius) comet or asteroid struck the Earth at that time, and that the many extinctions (including that of the dinosaurs) that occurred then were the result of some dramatic change in the environment caused by this cataclysm. The most-popular suggestion is that the upper stratosphere was filled with very fine dust to the point that no sunlight reached the Earth's surface for several months, and that all members of many species starved or froze during this period of darkness.

Meteorite Recovery

Until the past decade, meteorite recovery depended more on serendipity than planning. Meteorites were either found by people living in the area where an observed fall occurred or, in the absence of fall phenomena, were recognized by farmers, cowboys, prospectors, or rock hounds as being a rock type alien to the area.

Two recent developments that are enhancing meteorite recovery rates are (1) the establishment of photographic networks, and (2) the discovery of concentrations of meteorites in regions where the Antarctic ice sheet is eroding at unusually high rates. The photographic networks consist of arrays of cameras with rotating shutters that are photometrically triggered by bright fireballs. If a fireball is photographed by two stations, an exact trajectory can be established, and the fall site can be fixed to within a few kilometers. To date, the three photographic networks have each recorded one meteorite fall: the Příbram H5 chondrite by the Czech (now joined by the West Germans) All-Sky Network, the Lost City H5 chondrite by the now-defunct U.S. Prairie Network, and the Innisfree L6 chondrite by the Canadian Meteorite Observation and Recovery Project Network.[4] Figure I-9 shows the projection of the orbits of these objects on the ecliptic plane. The yield of new meteorites from these observing networks has been disappointing, but the orbital data are invaluable. The networks also yield precise orbital data for fireballs that fail to produce meteorite finds because the terminal mass is too small or because of unproductive ground searches.

Each orbit shown in Figure I-9 passes into or through the **asteroid belt** (shown stippled), the region between Mars and Jupiter that includes the orbits of all large and most small asteroids. This is an indication that the H-chondrite and L-chondrite parent bodies are (or were) located in the asteroid belt.

A large number of meteorites have been discovered in relatively rock-free areas of the great plains of North America and southwestern Australia. The efficient recovery of meteorites from such areas is partly a result of the educational efforts of H. H. Nininger, G. I. Huss, and other scientists willing to devote their time to field programs.

In 1969, Japanese scientists found nine meteorites of several different classes in an area of a few square kilometers on the Antarctic ice near the Yamato Mountains. During the 1974 and 1975 field seasons, another search party discovered 970 more meteorites in the same general area. The remarkable characteristic of this area is the blue color of the surface that indicates the presence of well-crystallized ice. The value of these cold-stor-

---

[4] Terminology used to describe meteorites is defined in Chapter II.

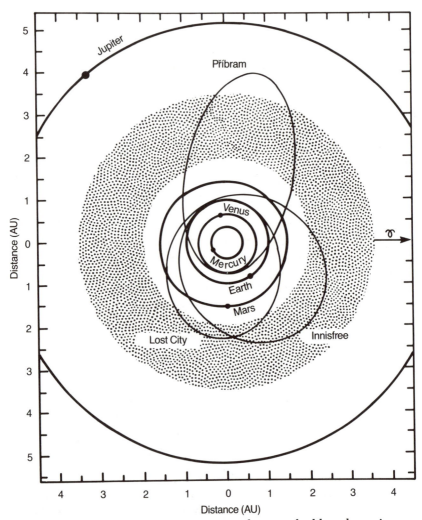

Figure I-9. The falls of three meteorites were photographed by telescopic cameras with rotating shutters. From the photographs, one can determine detailed atmospheric velocities and trajectories and, from this information, accurate orbits. The orbits generally lie in the region between the orbit of Venus and the outer reaches of the asteroid belt (stippled).

age meteorites is now generally recognized, and annual recovery expeditions to Antarctic are being financed by U.S., Japanese, and West German authorities. Terrestrial ages (see Chapter III) as great as ~0.7 Myr have been determined for some of the more weathered meteorites, whereas the fresh, unaltered appearance of others suggests much smaller ages. Because the temperature below the surface never exceeds 0°C, there is no melting and little mixing between the ice strata produced from the snow accumulated each season.

It is now understood that the blue ice is found in areas where the seaward flow of ice is hampered by mountains (many of them buried). Evaporation and scouring by high winds lead to net evaporation of the ice, exposing meteorites that fell at some other, more-inland location during the past million years. Although most, perhaps all, Antarctic meteorites have been altered by weathering, they are still well suited for many types of research. Their greatest value is in expanding the variety of meteoritic materials in our collections. Each new class of meteorites offers information about solar-system processes and/or locations not previously accessible.

One of the most-striking successes of the Antarctic meteorite-recovery program occurred in the 1981 recovery season with the discovery of a meteorite that almost certainly came from the Moon. It has long been known that a tiny fraction of the material ejected from lunar craters leaves the surface with velocities great enough to allow it to escape the Moon's gravitational field, and that 10 to 30 percent of such materials should eventually reach the Earth's surface. However, until the discovery of the new meteorite (designated Allan Hills A81005), no meteorite having properties closely related to recovered lunar materials was known. Properties indicating that the new stone is a regolith sample from the Moon are its bulk MnO/FeO ratio, oxygen-isotope compositions, cosmic-ray age, content of excess ^{40}Ar, and general mineralogical composition, including a high concentration of anorthite.

Micrometeorites

Small interplanetary dust particles can radiate their frictional energy away fast enough to survive atmospheric deceleration without melting. During the past few decades, attempts to collect these particles in the stratosphere with impactors or filters flown on airplanes, balloons, and rockets have been frustrated by contamination. In 1976, the first successful collections were reported by D. E. Brownlee and colleagues. Individual particles that had impacted onto a collecting plate carried to the stratosphere on a U-2 aircraft were examined in a scanning electron microscope. The researchers found a set of fine-grained particles that show no resemblance to known contaminants. Chondritic concentrations of nonvolatile elements were determined by electron-microscope studies, confirming the extraterrestrial nature of the particles. The particles are more fine-grained and friable than known meteorites, and it is speculated that they are of cometary origin. These particles represent an important new class of solar-system materials that are available for study in terrestrial laboratories, but their small size (from less than 0.1 μm to ~ 10 μm in diameter) will require the development of new techniques before trace constituents can be determined.

Suggested Reading

Buchwald, V. F. 1975. *Iron Meteorites.* University of California Press. In Volume 1 (of 3) are descriptions of the physics of atmospheric passage (Chapters 3 and 4), craters (Chapter 4) and fall statistics (Chapter 5).

King, E. A. 1976. *Space Geology.* Wiley. Chapter 1 is on meteorites, Chapter 2 on tektites, and Chapters 3 and 4 on craters.

Krinov, E. L. 1966. *Giant Meteorites.* Pergamon. Chapter 3 gives a detailed report on the fall of the Tunguska object.

Mason, B. 1962. *Meteorites.* Wiley. Chapters 1 and 2 give useful historical notes and descriptions of craters and observed falls.

Nininger, H. H. 1952. *Out of the Sky.* Dover. Numerous descriptions of fall phenomena and organized field-recovery programs by an innovative field expert.

Silver, L. T., and P. H. Schultz. 1982. *Geological Implications of Impacts of Large Asteroids and Comets on the Earth.* Special Paper **190**. Geological Society of America. A collection of technical papers presented at a conference stimulated by the evidence for a major impact at the end of the Cretaceous geological period.

Wetherill, G. W. 1974. Solar system sources of meteorites and large meteoroids. *Ann. Rev. Earth Planet Sci.* **2**:303. A technical discussion of evidence relating to meteorite orbits.

Wood, J. A. 1968. *Meteorites and the Origin of Planets.* McGraw-Hill. An introductory-level text. Chapter 1 describes fall phenomena, craters, and observing networks.

CHAPTER II

Composition and Taxonomy

A precise classification system is the key to understanding the biological or geological world; an imprecise system of classification leads to misinterpretations and confusion.

The development of every classificational system follows the same general evolutionary pattern. Properties of the members of the population are surveyed. Some are found to be discontinuous, and a tentative classification is based on these discontinuous properties. Additional surveys reveal other discontinuous properties that can be used to test the initial classification. Every classification scheme continues to evolve as more data become available. Interesting tensions develop between taxonomists who are mainly "lumpers" and those who are mainly "splitters," and between those researchers who are ready (sometimes too ready) to accept changes in a taxonomic system and those who cling to an "established" system long after it has been superceded. These tensions are well exposed at meetings of meteorite researchers.

Taxonomic Principles

The properties of meteorites are not continuous. Detailed studies of the mineralogy, chemical composition, or isotopic composition invariably divide the total population into clusters separated by unpopulated gaps in the distribution. Meteorites that fall together in these clusters in terms of a number of different properties are designated **groups**. For example, there are 13 groups of iron meteorites. Appendix A lists the recognized groups of meteorites. In the case of chondrites, it is sometimes advantageous to divide a group into smaller categories called **types**.

It is not accidental that the members of a group have similar properties. These similar properties result directly from the similar processes that formed the members of the group. Processes of meteorite formation differ from one solar-nebula location to the next and from one parent body to the next. The first step in understanding these processes is to determine which meteorites share a common origin. Then, by careful study of the members of that group, we can produce plausible models to account for their formation.

To simplify the classification, a **group** is required to have five or more members. Clusters consisting of one to four members also exist, but their inclusion in tables such as Appendix A would consume too much space. None of these **ungrouped** meteorites differs radically from the kinds of meteorites found in the groups. Although they tend to be ignored in sum-mary discussions such as this book, the ungrouped meteorites offer valuable clues regarding early solar-system history, and they are widely studied by meteorite researchers.

Chondrites

The stereotype image of a meteorite is a chunk of iron. In fact, however, iron meteorites account for only a small fraction (about 5 percent) of mete-orite falls. The most-common meteorites to fall are the chondrites, stony meteorites with compositions very similar to that of the Sun if the compari-son is restricted to nonvolatile elements. The greatest deviations from solar composition are depletions of the most-volatile elements.

In meteorite-research parlance, **abundance** is the atomic ratio of the element of interest relative to a normalizing element, typically silicon. The advantage of using abundances rather than **concentrations** (mass of element divided by mass of sample) is that abundances in diverse samples can be compared even if the total mass of one sample is not known (e.g., in the solar atmosphere) or if the samples differ in their content of some component such as H_2O or metal that may not be relevant to the desired compositional comparison.

In Figure II-1, calcium-normalized solar abundances determined by telescopic measurements of lines in the solar spectrum are compared with those of CI and CM chondrites, the most-volatile-rich classes of chondrites. The agreement between solar and CI abundances is good; the agreement between solar and CM chondrites is good for the nonvolatile elements, but abundances of three well-determined volatiles (Na, K, and Cd in Figure II-1) in CM chondrites are 30 to 50 percent lower than the solar values. The only other chondrite groups having volatile abundances comparable to CM (and lower than CI) levels are the EH and IAB chondrites, but these groups have Si/Ca ratios much higher than the reported solar ratio.

The determination of elements in the Sun is hampered by inadequate knowledge of the formation of spectral lines under the high-temperature (~ 6000 K) conditions prevailing in the solar photosphere, the visible part of the atmosphere. Because all elements can be determined more precisely in laboratory studies of meteorites, it is common practice to use CI-chon-drite abundance data as the best estimate of the composition of the Sun and of the primitive solar nebula from which the planetary portion of the solar system formed. Solar and CI abundances and CI concentration data are tabulated in Appendix D.

The name chondrite was originally based on the observation that such meteorites contain large amounts of small (0.1 to 2 mm) spheroids called

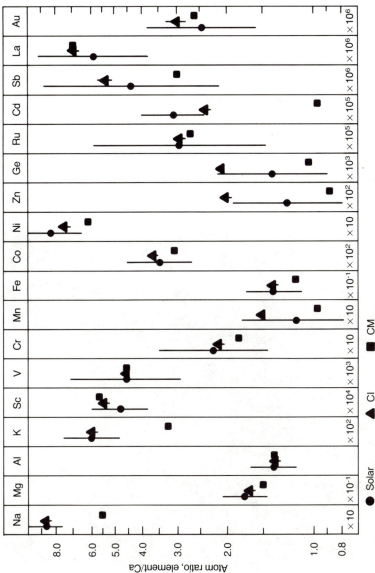

Figure II-1. Comparison of solar and chondritic element/Ca ratios for 18 elements moderately well determined in the solar atmosphere; the vertical bars show 70% confidence limits. Recent CM and CI chondrite abundance data of G. Kallemeyn and J. Wasson agree with the solar values for nonvolatile elements, but only CI abundances for the volatiles sodium, potassium, and cadmium are concordant with solar values. CI chondrites appear to be the materials most representative of mean solar-system composition.

chondrules (derived from the Greek word for grain; see Chapter VII for a more extensive discussion of chondrules and their origins). The Felix chondrite illustrated in Figure II-2 belongs to a group having a high concentration of chondrules and chondrule fragments. The name *chondrite* is now generally applied to any meteorite having solarlike composition, even though some (such as the volatile-rich CI chondrites just mentioned) are devoid of chondrules.

As discussed in more detail in Chapter VI, it appears that most extrasolar matter was vaporized by the conversion of gravitational to kinetic energy during the formation of the solar system. Gradual cooling of the resulting solar nebula resulted in the sequential condensation of elements in order of decreasing condensation temperatures. Those elements condensing before the common elements magnesium, silicon, and iron are designated **refractory** elements.

The abundance of refractory **lithophiles** (elements mainly found in sili-

a CM

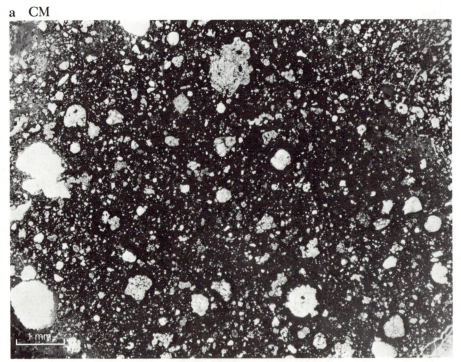

Figure II-2. Thin sections of *(a)* the Murchison CM chondrite and *(b)* the Felix CO chondrite, each photographed in transmitted light. The rounded objects are chondrules, which typically have diameters in the range 0.2 to 0.5 mm in the closely related CM and CO groups. The striking difference between the groups

cate minerals) is nearly constant within each chondrite group but varies considerably among the different groups and, as a result, is a useful parameter for classifying meteorites. Figure II-3 shows silicon-normalized abundances of three refractory lithophiles samarium, scandium, and aluminum plotted against a fourth, calcium. The nine plotted groups of chondrites tend to fall into five clusters on this diagram. Concentrations in a tenth group, the IAB chondrites, show considerable scatter, but their mean values fall near those in the L and H groups. Groups that have several properties in common comprise a clan; thus these data indicate five chondrite clans, and the IAB chondrites comprise a sixth. Table II-1 lists group abundances of the refractory elements aluminum and calcium and mean refractory abundances relative to those in CI chondrites.

The degree of oxidation of meteorites is closely related to the distribution of iron among its three common oxidation slates, 0, +2, and (more

b CO

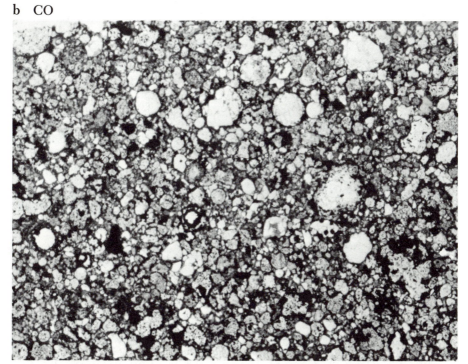

is in their abundance of fine-grained dark matrix, which accounts for ~50 percent of the CM mass, but only 20 to 30 percent of the CO mass. (Photos by W. R. Van Schmus)

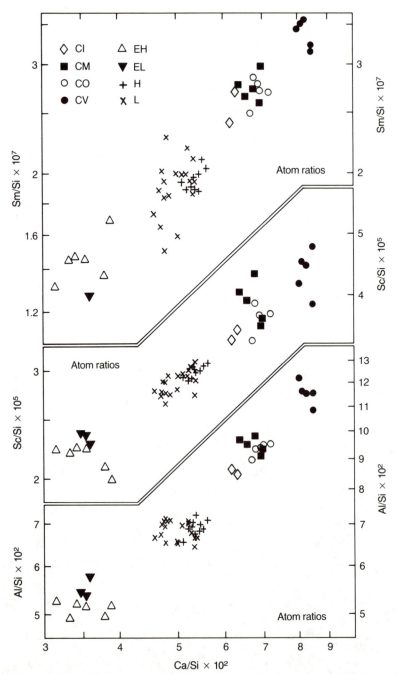

Figure II-3. Refractory-element abundances in the chondrite groups fall into five clusters; individual analyses by G. Kallemeyn are shown. This fractionation of refractories from silicon must have occurred during formation of the chondrites in the solar nebula, because textural evidence shows that no subsequent alteration has occurred. The observed variation makes refractory-element abundance a useful parameter for classifying the chondrites.

Table II-1
Properties of chondrite groups

Clan	Group	Al/Si (atom%)	Ca/Si (atom%)	refr/Si (norm*)	Fe/Si (atom%)	Fe/Si (norm*)	FeO/(FeO + MgO) (mol%)	Mafic-min. comp. (mol%)	Fe met/Si (atom%)	δ¹⁸O (‰)	Δ¹⁷O (‰)	Chondrule size†	Chondrule freq.‡
Refractory-rich	CV	11.6	8.4	1.35	76	0.87	35	§	0.6–19	1	−3	1.0	30
Minichondrule	CO	9.3	6.9	1.10	78	0.90	35	§	2.3–15	0	−4	0.2	70
	CM	9.7	6.8	1.13	80	0.93	43	§	0.1–0.5	7	−3	0.2	2
Volatile-rich	CI	8.6	6.2	1.00	86	1.00	45	§	<0.1	17	+1	—	<1
Ordinary	LL	6.5	4.7	0.76	53	0.62	27	Fa 26–32	2.7–22	4.9	1.3	0.47	80
	L	6.6	4.7	0.77	57	0.66	22	Fa 21–25	17–22	4.6	1.1	0.46	80
	H	6.8	4.9	0.79	80	0.93	17	Fa 17–19	46–52	4.2	0.8	0.33	80
IAB-inclusion	IAB	5–7	4–5	~0.7	~60	~0.70	6	Fs 4–8	~50	5.0	−0.4	—	—
Enstatite	EL	5.4	3.6	0.60	65	0.76	0.05	Fs ~0.05	47–57	5.7	0.0	0.5	—
	EH	5.1	3.6	0.59	97	1.13	0.05	§	68–72	5.7	0.0	0.5	20

* CI-normalized values.
† Median diameter in millimeters.
‡ Frequency as percentage. Metamorphism has destroyed the chondrule record in IAB and EL.
§ Mafic-mineral (olivine, Fa, or low-calcium pyroxene, Fs) compositions given only for equilibrated meteorites.

rarely) $+3$. For example, we can write a chemical reaction

$$Fe + \tfrac{1}{2}O_2 = FeO \qquad \text{(II-1)}$$

and for this reaction an equilibrium constant

$$K \simeq \frac{c_{FeO}}{c_{Fe}p_{O_2}^{1/2}} \qquad \text{(II-2)}$$

where c is concentration and p is partial pressure. We will let $p_{O_2}^{1/2}$, the square root of the partial pressure of O_2, be our measure of degree of oxidation. The concentration of Fe in Fe–Ni can generally be set equal to unity, and the concentration of FeO is approximately equal to the FeO/(FeO + MgO) ratio in the ferromagnesian minerals olivine and pyroxene (see mineral formulas in Appendix C). We then obtain the approximate relationship

$$\text{degree of oxidation} \equiv p_{O_2}^{1/2} \propto \frac{FeO}{FeO + MgO} \qquad \text{(II-3)}$$

This relationship is not valid for the CM and CI chondrites in which some iron is present in the $+3$ state, but it is clear that these groups have higher degrees of oxidation than the highest values observed in the other chondrites having negligible amounts of $+3$ iron.

The degree of oxidation is also a good classificational parameter; the FeO/(FeO + MgO) ratios listed in Table II-1 range from less than 0.1 to 45 mol%. The values assigned the CM and CI chondrites are their total FeO/(FeO + MgO) ratios but, as noted earlier, their degrees of oxidation are still higher than insertion of this value in Equation II-3 would indicate. The enstatite chondrites are so reduced that some silicon is present as the metal (dissolved in Fe–Ni). Also listed in Table II-1 are the FeO/(FeO + MgO) ratios observed in the mafic minerals olivine and pyroxene; Fa (for fayalite) indicates the ratio in olivine, Fs (for ferrosilite) indicates the ratio in pyroxene.

Another parameter that reflects the degree of oxidation is the fraction of iron present as metal. These values are listed in Table II-1. In Figure II-4, the abundance of reduced iron (defined as that in metal and FeS) is plotted against oxidized-iron abundance. Such a diagram gives a visual representation of both the bulk Fe/Si content (constant values lie along diagonal lines) and the degree of oxidation. Note that the six clans occupy distinct positions, although the separations in the carbonaceous-chondrite region are small.

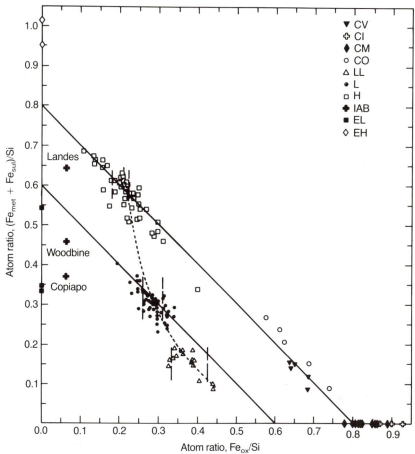

Figure II-4. Chondritic reduced iron (metallic iron and that present as FeS) is plotted against oxidized iron (that present in silicates and, in CM and CI, as Fe_3O_4). Lines having slopes of -1 correspond to constant total Fe/Si ratios; two are shown for reference purposes. A dashed curve depicts a suggested fractionation sequence linking the H (high-iron), L (low-iron) and LL (low-iron, low-metal) ordinary chondrites. The highly reduced enstatite (EH and EL) chondrites plot along the left axis, the highly oxidized CM and CI chondrites along the bottom axis. The fractionations of degree of oxidation and Fe/Si ratio occurred during the formation of these groups in the solar nebula.

Some closely related chondrites differ in their abundances of **sidero-philes**, elements that are concentrated in Fe–Ni if such phases (kamacite and taenite) are present. As discussed in Chapter VIII, these differences appear to result from differing efficiencies in the agglomeration of metal and silicates during planetesimal formation. The abundance of iron shown in Table II-1 is an example of these differences. Note that the groups comprising the enstatite and ordinary clans are resolved by this parameter.

Variations in the oxygen-isotopic composition are shown in Table II-1 as $\Delta^{17}O$, the difference in the $\delta^{17}O$ value from that expected on the basis of the observed $\delta^{18}O$ value and the 0.52 slope of the terrestrial fractionation line: $\Delta^{17}O = \delta^{17}O - 0.52\delta^{18}O$. See Chapter III for a discussion of oxygen-isotope systematics.

Textural differences are not easily included in a table, but two quantities (the mean abundance and size of chondrules) are included in Table II-1. Other textural and mineralogical features include the grain size, light reflectance, and presence of specified minerals.

As discussed in more detail in Chapters VII, VIII, and IX, the differences between the different chondrite groups almost certainly result from differences in nebular conditions at varying distance from the Sun.

Some chondrites are well equilibrated, with each grain of a given mineral (such as olivine or orthopyroxene) having essentially the same composition, and with the compositions of all minerals in the assemblage being essentially consistent with interequilibration of grains at a "metamorphic" temperature of ~ 1000 to ~ 1300 K. In other chondrites, individual minerals show wide ranges in composition, implying formation under very different conditions and little subsequent interequilibration. As discussed further in Chapters VII and VIII, these unequilibrated chondrites best preserve the record of the variety of conditions present in the primitive solar nebula.

It is useful to include in the classification symbol of each chondrite an indication of its degree of equilibration and metamorphic recrystallization. This is done by assigning the chondrites to **petrologic types** ranging from 1 in the least equilibrated to 6 (some workers use 7) in the most-equilibrated chondrites. These numbers are generally appended to the group symbol. Thus Allende is a CV3, Bruderheim an L6, and Abee an EH4 chondrite. Chondrite breccias are classified on the basis of their host fractions. In some ordinary-chondrite breccias, materials ranging from types 3 to 6 coexist.

Other parameters covary with the degree of equilibration and recrystallization. Volatile elements such as hydrogen, carbon, indium, bismuth, and the rare gases generally decrease in abundance with increasing petrologic type. Some components are absent in known type-1 chondrites (all of which are members of the CI group) and present in a systematically varying sequence in the other types. A number of the diagnostic properties are summarized in matrix form in Table II-2, a modified version of that originally proposed by Van Schmus and Wood.

It was earlier held that the least-equilibrated types, 1 and 2, best preserve the record of nebular processes. More recently it has been recognized that these types have been altered by hydrothermal processes which, for example, converted silicate minerals such as olivine to hydrated, clay-like

Table II-2
Criteria for distinguishing different petrologic types of chondrites

	Petrologic type					
	1	2	3	4	5	6
Homogeneity of olivine and pyroxene compositions	——	Mean deviations of pyroxene ≥5%, of olivine ≥3%		5% > mean pyroxene deviation >0%	Uniform ferromagnesian minerals	
Structural state of low-Ca pyroxene	——	Predominantly monoclinic		Abundant monoclinic crystals	Orthorhombic	
Degree of development of secondary feldspar	——	Absent		Predominantly as microcrystalline aggregates		Clear, interstitial grains
Igneous glass	——	Clear and isotropic primary glass, variable abundance		Turbid if present	Absent	
Metallic minerals	——	Taenite absent or very minor (Ni <200 mg/g)	Kamacite and taenite (Ni >200 mg/g) present			
Mean Ni content of sulfide minerals	——	>5 mg/g	<5 mg/g			
Overall texture	No chondrules	Very sharply defined chondrules		Well-defined chondrules	Chondrules readily delineated	Poorly defined chondrules
Texture of matrix	All fine-grained, opaque	Much opaque matrix	Opaque matrix	Transparent microcrystalline matrix	Recrystallized matrix	
Bulk carbon content	30–50 mg/g	8–26 mg/g	2–10 mg/g	<2 mg/g		
Bulk water content	180–220 mg/g	20–160 mg/g	3–30 mg/g	<15 mg/g		

Source: R. Van Schmus and J. Wood, *Geochim. Cosmochim. Acta* 31:747 (1967) and more recent results.

minerals. The nebular record is, in fact, best preserved in the least equilibrated type-3 chondrites, as found in groups CO, CV, H, L, LL, and EH.

Differentiated Meteorites

The following scenario appears to be valid for the Earth, and it probably also is valid for the parent planets of some groups of differentiated meteorites. The planet's composition was essentially chondritic, similar to one of the metal-bearing chondrite classes. Complete melting of the interior of such a planet led to the separation of immiscible metal and silicate. The denser metal (the density of solid metal is $\rho = \sim 7.9$ g $\cdot$ cm^{-3}; that of a metallic liquid is somewhat smaller depending on its S content) migrated to the center and formed a core. Further differentiation of the silicates ($\rho = 2.6$ to 3.6 g $\cdot$ cm^{-3}) led to the extrusion of a lighter, fusible silicate melt on the surface, whereas denser and more-refractory silicates remained behind to form a mantle. The differentiated meteorite groups can be fit into such a picture, but it is oversimplified, and a caveat is necessary: each group is stamped by the details of its formation, which often involved multiple melting or mixing events. Properties of differentiated silicate-rich meteorites are listed in Table II-3.

Silicate-rich differentiated meteorites having low metal contents are often designated **achondrites.** The two groups having high metal contents are often called **stony-irons.** The problem with these terms is that they tend to take on unjustified generic significance. For example, the mesosiderite stony-irons and the howardite achondrites are much more closely related than are certain achondrite groups (for example, the howardites and the ureilites). I recommend minimizing the use of the terms achondrite and stony-iron. For mnemonic purposes, it is useful to note that any group given a name that does not include "chondrite" or "iron" is a differentiated silicate-rich meteorite.

On Earth, melting of the mantle generally produces **basalts,** rocks composed of roughly equal amounts of plagioclase and pyroxene. The **eucrites** are a group of basaltic meteorites. Figure II-5 shows a photograph of Ibitira, one of the few eucrites that escaped crushing. The Earth's mantle is composed primarily of ultramafic rocks — rocks having high contents of the mafic ("dark") minerals olivine and (to a lesser extent) pyroxene. The **diogenites** are meteorites having ultramafic compositions. They were previously held to represent the mantle of an asteroid, but alternative proposals involving smaller magmatic systems have recently been proposed. Most eucrites and diogenites have been crushed to form **breccias,** rocks formed from the fragments of preexisting rocks. Because the individual fragments of eucrite and diogenite breccias are from the same rock type, these breccias are designated **monomict.** In some chondrite breccias, the

Table II-3
Properties of differentiated silicate-rich meteorites

| Group | Major mafic mineral | | | Fe–Ni (mg/g) | $\delta^{18}O$ (‰) | $\Delta^{17}O$ (‰) | Breccia type |
	Name	Concentration°	FeO/(FeO + MgO) (mol%)				
Eucrites†	Pigeonite	400–800	45–70	<10	3.2–3.8	−0.2	Monomict
Howardites†	Orthopyroxene	400–800	25–40	~10	3.2–3.8	−0.2	Polymict
Diogenites†	Orthopyroxene	~950	25–27	<10	3.2–3.8	−0.2	Monomict
Mesosiderites‡	Orthopyroxene	400–800	23–27	300–550	3.2–3.8	−0.2	Polymict
Pallasites†	Olivine	~980	11–14	280–880	3.2–3.8	−0.3	Monomict‡
Aubrites	Low-calcium clinopyroxene	~970	0.01–0.03	~10	5.1–5.5	0.0	Polymict
Ureilites	Olivine	~850	10–25	10–60	7.6–8.0	−1.0	Polymict

° As fraction of silicates.
† Together these five groups comprise the igneous clan.
‡ Pallastic silicates monomict; metal and silicates probably originally separate in core and mantle, respectively.

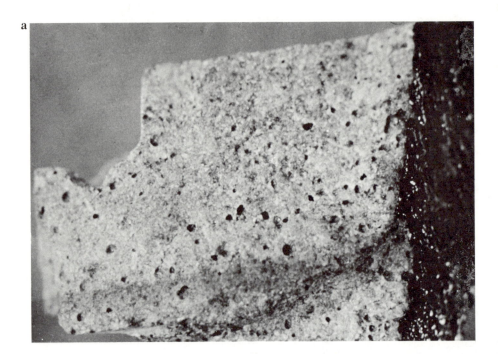

Figure II-5. The unbrecciated Ibitira eucrite shows vesicles similar to those commonly seen in chilled terrestrial lavas. The vesicles resulted from gas formation following extrusion of the basaltic lava into a low-pressure, surficial environment. Before extrusion, the volatile phase was dissolved in the silicate melt. (a) The glossy black fusion crust characteristic of eucrites is visible at the top. (b) The face is about 2.0 × 2.3 cm. (Photos by F. Wlotzka and P. Deibele)

differences among clasts are only in degree of metamorphic reheating; these breccias are designated **genomict**. Breccias consisting of two or more rock types are designated **polymict**. The **howardites** are polymict breccias containing several rock types but consisting dominantly of rocks similar to the eucrites and the diogenites.

The silicates in the polymict **mesosiderites** are very similar to those in the howardites; the differences are resolvable only by detailed petrological or chemical studies. These curious objects appear to be regolith breccias. The metal may have originated when the surface of an asteroid was impacted by a projectile consisting of a core or core fragment of a fragmented asteroid. This metal seems to have accreted at very low velocity (less than 1 km · s^{-1}), because high-speed impacts lead to low projectile/target ratios in the crater ejecta. Alternative suggestions are that the silicates were the projectile and the metal was exposed on the surface of a large body, or that the metal—silicate ratio in the mesosiderites is not typical of the entire crater ejecta.

The **pallasites** consist of roughly equal portions of metal and the ultramafic mineral olivine. Olivine is probably the dominant mineral in the Earth's mantle, and it is generally accepted that most pallasites originated at the interface between an olivine lower mantle and a metallic core. Three ungrouped pallasites called the Eagle-Station trio have unusual compositions. It has been suggested that they formed from a melt generated by a large impact event on the surface of an undifferentiated parent body.

The oxygen-isotope composition of the eucrites, diogenites, howardites, mesosiderites, and pallasites are very similar (Table II-3). As discussed in more detail in Chapter V, their elemental and phase compositions also appear to be consistent with their joint formation in a single parent body. To emphasize this close relationship, I designated these five groups the **igneous clan**.

The **aubrites** consist almost entirely of the pyroxene enstatite ($MgSiO_3$); its very low iron content indicates an extremely low degree of oxidation, and this is confirmed by the presence of up to 1.2 percent of metallic silicon dissolved in the rare Fe–Ni grains. The oxygen-isotope compositions of aubrites are essentially identical to those in EH and EL chondrites, and the highest metallic-silicon concentrations are identical to those found in the EL chondrites. A close genetic link with the enstatite clan chondrites is thus inferred.

The **ureilites** are curious objects consisting of mixtures of ultramafic silicates with a carbon-rich material that is rich in planetary-type rare gases and thus is apparently primitive in nature (see Chapter III). Diamonds of shock origin are found in the carbon-rich component. Ureilite oxygen-isotope compositions are somewhat similar to those of CM chondrites, well removed from those of other groups of differentiated meteorites.

The ungrouped differentiated silicate-rich meteorites are generally simi-

lar to those just described, but they record distinct differences in formational conditions on other parent bodies. Among these are the shergotty –
nakhla – chassigny clan, which shows volatile abundances and isotopic
ratios consistent with formation on the surface of Mars. (As discussed in
Chapter V, there are some difficulties with this hypothesis of origin.) Information regarding other ungrouped achondrites can be found in more detailed treatises.

Polishing and etching a planar section of an iron meteorite generally
reveals a striking structure. Irons consisting entirely of single crystals of
kamacite (the low-nickel α-iron phase) are called **hexahedrites;** their structures are marked only by the presence of occasional inclusions of FeS or
$(Fe,Ni)_3P$ and oriented sets of lines (Neumann lines) of shock origin (Figure
I-3). The name *hexahedrite* refers to the fact that these single crystals have
cubic (regular hexahedron) crystalline structures. The nickel concentration
in hexahedrites is always below 58 mg/g and only rarely below 53 mg/g.

Most irons are **octahedrites** formed by precipitation at 500 to 700°C of
kamacite lamellae along four sets of crystalline planes of the high-temperature taenite (γ-iron) phase. The sections through these lamellae exposed on
a planar surface are commonly called kamacite bands. This formation,
which requires low cooling rates, is discussed in more detail in Chapter IV.
The angles between the four sets of planes are identical to those between
the faces of a regular octahedron (note that there are only four face orientations in an octahedron because opposite faces are parallel). The thickness to
which kamacite lamellae grew depended on the nickel concentration and
the parent-body cooling rate: the lower the Ni concentration and/or the
lower the cooling rate, the thicker the lamellae. The mean thickness
(loosely called the "bandwidth") may range from less than 0.1 mm to several centimeters. To simplify structural descriptions, octahedrites are divided into a bandwidth sequence consisting of finest (less than 0.2 mm),
fine (0.2 to 0.5 mm), medium (0.5 to 1.3 mm), coarse (1.3 to 3.3 mm) and
coarsest (greater than 3.3 mm) octahedrites. In the older literature, these
bandwidth ranges are sometimes treated as classes, but compositional data
show that bandwidth measurements alone do not lead to the grouping of
genetically related irons.

Figure II-6 shows the structure of the Osseo IAB coarse octahedrite.
Such a structure is revealed by polishing a planar surface and etching it with
a very dilute solution of nitric acid in ethyl alcohol. The plane of this section
is essentially parallel to one of the octahedral planes, which shows up as
irregular blotches on the surface. Intersections of lamellae having the other
three possible orientations form 60° angles on this surface. The roundish
FeS + C inclusions surrounded by $(Fe,Ni)_3P$ are characteristic of group IAB
and IIICD irons. They are not found in coarse octahedrite members of other
groups (for example, in IIIE irons).

Figure II-7 shows the structure of a large section of the Gibeon IVA fine

0 1 2 3 4 5 cm

Figure II-6. The Osseo IAB coarse octahedrite consists almost entirely of
kamacite bands. This photo made with oblique lighting brings out the contrast
in reflectivity even among bands having the same octahedral orientation.
Several rounded inclusions consist of medium-gray troilite (FeS) and black
graphite (C) surrounded irregularly by light-gray schreibersite ((Fe,Ni)$_3$P). In
the upper center, a preterrestrial crack (perhaps of shock origin) has been filled
with terrestrial oxidation products. (Smithsonian Institution Photo)

octahedrite. The only inclusions visible in this photo consist of FeS. This
section shows discontinuities in the orientation of the **Widmanstätten pattern** (another name for the octahedral structure). The differences reflect
the fact that, prior to kamacite precipitation, three distinct taenite crystals
occupied these areas; this process is discussed further in Chapter V.

Finely divided mixtures of kamacite and taenite having bulk nickel concentrations greater than ~120 mg/g are called **plessite**. Irons that consist
mainly of fine plessite but that also show a few very fine kamacite spindles
are designated **plessitic octahedrites**. Figure II-8 shows the Butler plessitic
octahedrite, an ungrouped iron. The distinction between finest octahedrites and plessitic octahedrites is that the former have long, continuous

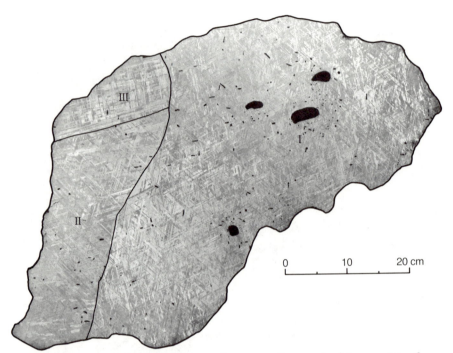

Figure II-7. This large section of the Gibeon IVA fine octahedrite consists of three regions (I, II, and III) having different octahedral orientations of the kamacite lamellae. The dark inclusions are FeS (troilite). Gibeon fell as a gigantic, prehistoric shower in Namibia (South-West Africa). (Photo from R. Schaudy et al., *Icarus* 17:174, 1972)

bands, whereas bands in the latter tend to consist of isolated spindles or intersecting "sparks ."

Meteorites whose metallic structure consists entirely of plessite (perhaps with a few rare kamacite prisms) are called **ataxites,** meaning "without structure." Figure II-9 shows an ataxite structure; under careful sample preparation and skilled microscopy, many structural details are visible. The malaprop name dates back to a time when the available microscopes were not adequate to resolve the structure.

Until a decade ago, iron meteorites were classified primarily on the basis of their bandwidths. However, detailed trace-element studies on a large set of iron meteorites revealed that there is only a very rough correlation between structure and compositional groupings. The modern classfication of iron meteorites includes detailed structural study (not just bandwidth measurements) and treats this as information complementary to the chemical composition.

Figure II-8. The ungrouped Butler iron has a plessitic octahedrite structure; kamacite (α-iron) is oriented along the octahedral directions (see text) but forms spindles and sparks rather than long continuous bands. Taenite (γ-iron) is present as darker-gray borders on the lighter-gray kamacite. Between the kamacite "bands" is dark plessite having no visible structure next to the taenite but having tiny, oriented kamacite sparklets in the central regions well away from taenite. The long dimension of the photograph is about 10 mm. (Photo by J. I. Goldstein)

Figure II-9. High magnification shows that the Tlactopec IVB iron consists mainly of very fine plessite (the lighter blebs are tiny kamacite sparklets). A few rare "large" prisms of kamacite surrounded by taenite, such as the two shown here, are occasionally found. The designation ataxite (meaning "structureless") was assigned at a time when high-power microscopy was not possible. The long dimension of the photograph is $\sim 340\ \mu m$. (Photo by V. F. Buchwald)

The two trace elements, gallium and germanium have proven to be iron-meteorite classificatory parameters *sans pareil*. In Figure II-10 we see that, with the exception of groups IAB and IIICD, gallium and germanium show narrow ranges (factors of less than 1.6) within groups but cover ranges of 2000 and 40,000, respectively, within the iron meteorites as a whole. In general, knowledge of gallium or germanium content, nickel content, and the structure suffices to classify most iron meteorites into one of 13 groups, each having five or more members. About 13 percent of the irons are "ungrouped"—in other words, they form sets of one to four compositionally related irons.

In contrast to gallium and germanium, iridium tends to show very wide ranges (factors of up to 5000) within groups (Figure II-11). As a result, determination of iridium serves to distinguish an iron meteorite from about

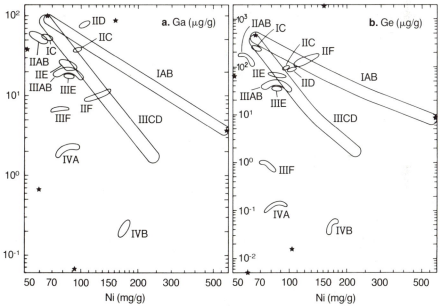

Figure II-10. As shown on these plots of *(a)* gallium versus nickel and *(b)* germanium versus nickel, the ranges of gallium and germanium concentrations tend to be small in iron-meteorite groups relative to the total ranges among all irons (note the different scales on the logarithmic axes). As a result, gallium and germanium concentrations are very useful parameters for classifying the iron meteorites. The outlined areas on this diagram show the locations of the 13 iron-meteorite groups. Points representing the individual meteorites (including the ~13 percent that are ungrouped) are not shown. The stars show the locations of the six meteorites having extreme nickel, gallium, and germanium concentrations.

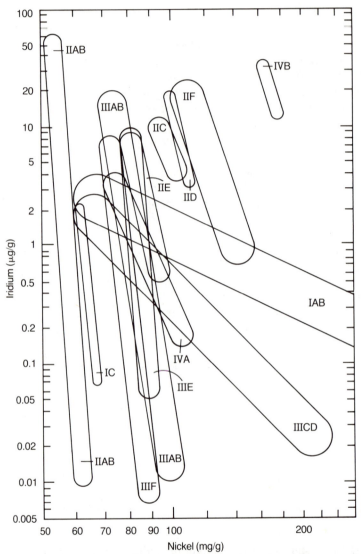

Figure II-11. The concentration of Ir shows large ranges (over factors as large as 6000) in iron-meteorite groups. Because of this large range, iridium concentration is a valuable parameter for determining whether or not two members of a group are fragments from the same fall or are otherwise closely related.

Table II-4
Properties of 13 iron-meteorite and two other metal-rich groups

Group	Freq. (%)	Bandwidth (mm)	Structure[a]	Ni (mg/g)	Ga (μg/g)	Ge (μg/g)	Ir (μg/g)	Ge–Ni correlation[§]
{ IA	17.0	1.0–3	Om–Ogg	64–87	55–100	190–520	0.6–5.5	neg }
IB	1.7	0.01–1.0	D–Om	87–250	11–55	25–190	0.3–2.0	neg }
IC	2.1	<3	Anom, Og	61–68	49–55	212–247	0.07–2.1	neg
{ IIA	8.1	>50	H	53–57	57–62	170–185	2–60	pos? }
IIB	2.7	5–15	Ogg	57–64	46–59	107–183	0.01–0.9	neg }
IIC	1.4	0.06–0.07	Opl	93–115	37–39	88–114	4–11	pos
IID	2.7	0.4–0.8	Of–Om	96–113	70–83	82–98	3.5–18	pos
IIE	2.5	0.7–2	Anom†	75–97	21–28	62–75	1–8	abs
IIF	1.0	0.05–0.21	D–Of	106–140	9–12	99–193	0.8–23	pos
MES	—	ca. 1	Anom	61–101	8.9–16	37–56	2.2–6.2	abs
PAL	—	ca. 0.9	Om	79–129	14–27	29–71	0.01–2	neg?
{ IIIA	24.8	0.9–1.3	Om	71–93	17–23	32–47	0.15–20	pos }
IIIB	7.5	0.6–1.3	Om	84–105	16–21	27–46	0.01–0.15	neg }
{ IIIC	1.4	0.2–3	Off–Ogg	62–130	11–92	8–380	0.07–2.1	neg }
IIID	1.0	0.01–0.05	D–Off	160–230	1.5–5.2	1.4–4.0	0.02–0.07	neg }
IIIE	1.7	1.3–1.6	Og	82–90	17–19	34–37	0.05–6	abs
IIIF	1.0	0.5–1.5	Om–Og‡	68–85	6.3–7.2	0.7–1.1	0.006–7.9	neg
IVA	8.3	0.25–0.45	Of	74–94	1.6–2.4	0.09–0.14	0.4–4	pos
IVB	2.3	0.006–0.03	D	160–180	0.17–0.27	0.03–0.07	13–38	pos

Note: Brackets combine compositionally distinct portions of single groups.
[a] Structure abbreviations: H, hexahedrite; Ogg, Og, Om, Of, Off, Opl, coarsest, coarse, medium, fine, finest, and plessitic octahedrites; D, ataxite; Anom, anomalous structure that does not fit into the other categories.
† Also Om and Ogg.
‡ Also Ogg and Of.
§Abbreviations: *positive*, *negative*, *absent*.

98 percent of the other irons in the same group. This is very useful when one wishes to determine whether two iron meteorites found within the same area are fragments from the same fall. Table II-4 summarizes the properties of iron meteorites.

Near the beginning of the discussion on iron-meteorite classification I noted that earlier (as recently as two decades ago) the iron meteorites were classified on the basis of bandwidth, but that this did not result in resolved genetically related groups. On the other hand, now that groups have been defined on the basis of compositional and structural evidence, it is possible for an expert to classify a large fraction (perhaps 70 percent) of the iron meteorites on the basis of structural information alone. For example, all single-crystal hexahedrites (Figure I-4) are members of the IIA portion of groups IIAB; all irons with the metal-inclusion texture closely similar to that shown in Figure II-6 are members of group IAB or the closely related minor group IIICD. Fine octahedrites having textures like that of Gibeon (Figure II-7) that are free of inclusions except for minor FeS nodules are almost always members of group IVA.

Compositional studies of the metallic phases of silicate-rich differentiated meteorites also provide valuable classificational information. In this way it has been shown that pallasites divide into one large "main group" and at least two additional compositional clusters. Properties of the metal of mesosiderites and main-group pallasites are also listed in Table II-4.

What is the significance of these groups that have been defined? There is abundant evidence that different groups of the same rock type (different groups of chondrites or irons) were formed in different parent bodies and generally at differing distances from the Sun. In contrast, there exists the possibility that some of the different types of differentiated meteorites (possibly all the members of the igneous clan) formed in the same parent body. These points are discussed in more detail in later chapters.

Suggested Reading

Clayton, R. N., N. Onuma, and T. K. Mayeda. 1976. A classification of meteorites based on oxygen isotopes. *Earth Planet. Sci. Lett.* **30**:10. A technical discussion of the application of oxygen-isotope data to the classification of meteorites.

Dodd, R. T. 1981. *Meteorites: A Petrologic-Chemical Synthesis.* Cambridge University Press. 368 pp. Chondrite classification is discussed in Chapter 2, differentiated-meteorite classification in Chapter 7.

Hutchison, R. 1983. *The Search for Our Beginning.* Oxford University Press. 164 pp. There is much general information about meteorites in this book. Chapter 4 gives a good overview of classification; color plates 3 and 4 show textures of a variety of kinds of meteorites.

Mason, B. 1962. *Meteorites.* Wiley. 274 pp. Chapters 7 and 8 give a good but dated description of the composition and classification of differentiated silicate-rich meteorites.

Scott, E. R. D., and J. T. Wasson. 1975. Classification and properties of iron meteorites. *Rev. Geophys. Space Phys.* **15**:527. The classification of iron meteorites is discussed in detail.

Wasson, J. T. 1974. *Meteorites: Classification and Properties.* Springer. 316 pp. Chapter 2 gives a detailed discussion of meteorite classification.

CHAPTER III

Age, Isotopes, and Interstellar Grains

Each chemical element consists of several **isotopes,** atomic species having the same number of protons and thus the same nuclear charge, but differing in their contents of neutrons and thus in their mass numbers. Some isotopes are stable, others are radioactive. An isotope is commonly written symbolically as ^{M}X, where X is the symbol of the chemical element and M is the mass number of the isotope. For example, oxygen has three stable isotopes: ^{16}O, ^{17}O, and ^{18}O. The generic term **nuclide** refers to any atomic species defined by given numbers of protons and neutrons in the nucleus; thus ^{12}C, ^{14}N, and ^{16}O are different nuclides.

Variations in the isotopic composition of meteoritic elements can result from any one of four main causes: (1) formation of isotopes by decay of a radioisotope; (2) formation of isotopes as a result of nuclear reactions produced by cosmic rays; (3) fractionation of isotopes resulting from mass-dependent differences in kinetic (rate-dependent) or thermodynamic (energy-dependent) properties; (4) mixing of materials that have different isotopic compositions because of differences in element-formation processes in stars or in interstellar chemical reactions.

Such isotopic data are exceedingly important for establishing many key facts about the solar system and the processes that led to the formation of the planets, including the following.

1. The age of the solar system, based on ^{87}Rb – ^{87}Sr dating is 4.53 Gyr, with an uncertainty of perhaps 0.02 Gyr; other isotope systems yield similar ages.
2. The different groups of chondritic meteorites were formed within a period of 16 Myr or less at the beginning of solar-system history.
3. Certain meteorite groups (the chondrite H group, and iron meteorite groups IIIAB and IVA) experienced major breakups that generated a large fraction of the members of these groups in a single event.
4. Most elements in most meteorite groups are identical in isotopic composition; thus nebular materials must have been relatively well mixed in interstellar or nebular processes.
5. Oxygen isotopic ratios show large variations within primitive chondrites and between groups of meteorites; these data imply distinct prenebular reservoirs that were incompletely mixed during nebular processes.
6. The isotopic compositions of rare-gas (He, Ne, Ar, Kr, Xe) or light (H, C, N. O, S) elements can be used to infer the nature of the star in which these elements were synthesized.

Radiometric ages

The **halflife** $T_{1/2}$ of a radionuclide is the time during which one-half of the amount initially present decays. The decay constant λ is related to the halflife by

$$\lambda = \frac{0.693}{T_{1/2}} \tag{III-1}$$

where 0.693 is the natural logarithm of 2. Halflives and decay constants for several common radionuclides are listed in Appendix E.

   If we designate the amount of a radioactive parent nuclide present at time t as P and the amount initially present (at $t = 0$) as P_I, then the radioactive decay law can be written as

$$P = P_I e^{-\lambda t} \tag{III-2}$$

or

$$P_I = P e^{\lambda t} \tag{III-3}$$

The amount D of a daughter nuclide present at time t is given by

$$D = D_I + (P_I - P) \tag{III-4}$$

or

$$D = D_I + P(e^{\lambda t} - 1) \tag{III-5}$$

where D_I is the amount of the daughter present initially. If the parent decays to more than one daughter (that is, ^{40}K decays either to ^{40}Ar or ^{40}Ca), then one must multiply the second term by a factor α equal to the fraction of the decays that produce the daughter D:

$$D = D_I + \alpha P(e^{\lambda t} - 1) \tag{III-6}$$

For example, $\alpha = 0.12$ for $^{40}K \rightarrow {}^{40}Ar$. Because the values of α and λ are well determined for the relevant radionuclides, determination of D, D_I, and P allows the calculation of an age t.

In almost every case it is a high-temperature event that establishes the time $t = 0$ (starts the radiometric clock running). In the case of rare-gas daughters, the high-temperature event serves to expel any daughter nuclide remaining from the previous decay of the parent. In cases where the daughter is not a rare gas, the high-temperature event produces isotopic equilibration among all the mineral phases of the rock.

We can distinguish several types of radiometric ages. **Formation ages** measure the time elapsed between the first resolvable isotopic equilibration event and the present. If later reheating events have led to partial isotopic equilibration, then the resulting age is called a **metamorphism age**. The distinction between these two types of ages disappears as the degree of reequilibration during the reheating event increases.

In very old meteorites, another type of age called a **formation interval** can be determined. Small amounts of radionuclides with halflives in the range of 1 to 100 Myr were still present when the meteorites formed, but they have decayed to unmeasurably low levels during the intervening 4.5 Gyr. The former presence of these extinct radionuclides is indicated by measurable amounts of their daughters. The prime example is the decay $^{129}I \rightarrow {}^{129}Xe$ with halflife of 16 Myr.

The most-precise ages result from the mass-spectrometric determination of the isotopic composition and concentration of the daughter and the parent concentration. Formation ages are generally based on the following parent–daughter pairs: $^{40}K \rightarrow {}^{40}Ar$; $^{87}Rb \rightarrow {}^{87}Sr$; $^{147}Sm \rightarrow {}^{143}Nd$; $^{232}Th \rightarrow {}^{208}Pb$; $^{235}U \rightarrow {}^{207}Pb$; $^{238}U \rightarrow {}^{206}Pb$. Because the last three decays all produce isotopes of lead, it is common practice to combine them into a single study, where the redundancy provides additional control on the quality of the age determinations. Because of space limitations, we limit our discussion here to the Rb–Sr and K–Ar methods.

The nuclide ^{87}Rb decays entirely to ^{87}Sr, and thus α is unity. Substituting into Equation III-6, we then obtain

$$^{87}Sr = {}^{87}Sr_I + {}^{87}Rb(e^{\lambda_{87}t} - 1) \tag{III-7}$$

Because the amount $^{87}Sr_I$ is directly proportional to the amount of nonradiogenic strontium isotopes, it is convenient to divide this equation by the amount of ^{86}Sr, an abundant strontium isotope that is not produced by radioactive decay:

$$\frac{^{87}Sr}{^{86}Sr} = \left(\frac{^{87}Sr}{^{86}Sr}\right)_I + \frac{^{87}Rb}{^{86}Sr}(e^{\lambda_{87}t} - 1) \tag{III-8}$$

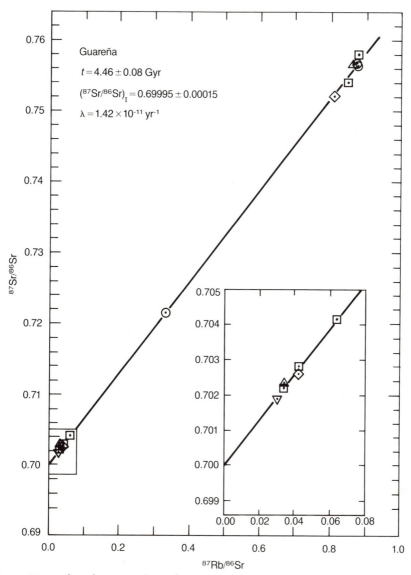

Figure III-1. The Rb–Sr isochron for the Guareña H6 chondrites. Most points represent data on mineral fractions separated according to density. The slope of a least-squares best-fit line yields an age of 4.46 Gyr. The insert shows more details of the points having the lowest Rb–Sr ratios. (From G. J. Wasserburg et al., *Earth Planet. Sci. Lett.* 7:33, 1969)

The two unknowns for each rock are its age t and its initial isotope ratio $(^{87}Sr/^{86}Sr)_I$. Measurement of $^{87}Sr/^{86}Sr$ and $^{87}Rb/^{86}Sr$ in two fractions of the rock differing in $^{87}Rb/^{86}Sr$ ratio allows us to solve for these two unknowns. The greater the number of fractions and the more different their $^{87}Rb/^{86}Sr$ ratios, the greater the precision with which the unknown quantities can be determined.

Fractions having different $^{87}Rb/^{86}Sr$ ratios are obtained by mineral separations. For example, ratios are generally high in plagioclase but low in pyroxene. Figures III-1 and III-2 show mineral isochrons for the meteorites Guareña (an H6 chondrite) and Ibitira (an unbrecciated eucrite). Both meteorites were formed very early in solar-system history. The higher uncertainty associated with the Ibitira age results chiefly from its lower range of $^{87}Rb/^{86}Sr$ ratios. The separation of mineral fractions without con-

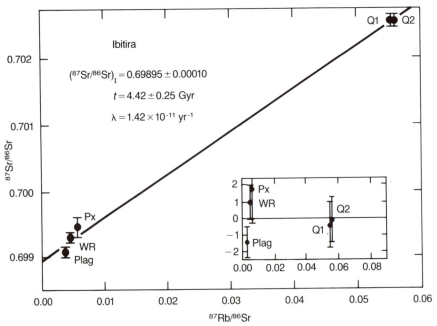

Figure III-2. The Rb–Sr isochron for the Ibitira eucrite. The large age uncertainty based on the least-squares best-fit slope results from the very small range in $^{87}Sr/^{86}Sr$ ratios among the different mineral fractions (compare with Figure III-1). The Q fractions are fine-grained materials interstitial to the dominant pyroxene and plagioclase; WR stands for whole rock (the unseparated bulk sample). The inset shows the uncertainties in ϵ, the difference between the measured $^{87}Sr/^{86}Sr$ ratio and that given by the isochronic line. (From J. L. Birck and C. J. Allegre, *Earth Planet. Sci. Lett.* 39:37, 1978)

tamination or alteration is a major challenge. In these studies, density fractions were separated by sinking or floating in liquids of differing density.

Table III-1 lists Rb–Sr ages and initial $^{87}Sr/^{86}Sr$ ratios. The most precise chondrite ages cluster in the range 4.45 to 4.53 Gyr. The lower of these values may reflect some resetting during metamorphism; the higher values offer the best estimate for the age of the solar system, because differences in formation intervals (discussed below) are only of the order of a few tens of megayears. I suggest using 4.53 ± 0.02 Gyr as the best estimate of the mean age of the solar system.

It is apparent from Table III-1 that a large fraction of the differentiated meteorites also have ages of at least 4.5 Gyr. Apparently, extensive heating and melting of the parent planets occurred within the first 100 Myr of solar-system history. As discussed further in Chapter IV, the heat source cannot yet be defined.

The low Kodaikanal and Kapoeta ages probably reflect severe heating, including melting of the Kodaikanal material as a result of impacts. Note that one Kapoeta clast seems to have escaped this later heating. The ungrouped Nakhla and two similar stones have formation ages near 1.3 Gyr. The event that started their clocks may have been either igneous or metamorphic in nature; petrographic data do not allow a clearcut decision.

Because the daughter ^{40}Ar produced by ^{40}K decay is a rare gas, it is partially lost by diffusion from certain "leaky" minerals, even in the absence of reheating events. As a result, whole-rock $^{40}K - ^{40}Ar$ ages tend to be lower than the true formation ages. One solution to the problem would be to determine mineral isochrons analogous to those described for the $^{87}Rb - ^{87}Sr$ system, and then to base the slope only on the minerals having high $^{40}Ar/^{40}K$ ratios. An alternative widely used at present involves selective outgassing of mineral fractions without mechanical separation.

In this method, commonly called the $^{39}Ar - ^{40}Ar$ method, the sample is first placed in a neutron flux in a nuclear reactor, where the more abundant stable isotope ^{39}K is converted to radioactive but long-lived ^{39}Ar by the gain of a neutron and loss of a proton. At the same time, a portion of the most abundant calcium isotope ^{40}Ca is converted to long-lived radioactive ^{37}Ar by gain of a proton and loss of an alpha particle (a ^4He nucleus). Neutron-flux monitors consisting of known amounts of potassium and calcium are included in the irradiation package. The sample is returned to the laboratory, and argon is extracted in a series of temperature steps, typically about 1 h per step at temperature intervals of 100°C to 200°C. The amounts of K, Ca, and ^{40}Ar released during each step are determined by mass-spectrometric measurements of ^{39}Ar, ^{37}Ar, and ^{40}Ar, respectively; the stable nonradiogenic argon isotopes ^{36}Ar and ^{38}Ar also are measured.

After data reduction, the results are commonly presented as shown in

Table III-1
The Rb–Sr "internal-isochron" formation ages and initial $^{87}Sr/^{86}Sr$ ratios of meteorites

Meteorite	Class	Age° (Gyr)	$(^{87}Sr/^{86}Sr)_I$	Reference (see notes)
Chondrites				
Indarch	EH4	4.39 ± 0.04	0.7005 ± 0.0009	13
Indarch	EH4	4.46 ± 0.08	0.7005 ± 0.0017	6
Saint Mark's	EH5	4.34 ± 0.05	0.69979 ± 0.00022	13
Saint Sauveur	EH5	4.46 ± 0.05	0.6993 ± 0.0014	13
Campo del Cielo	IA	4.6 ± 0.1	0.699	19
Guareña	H6	4.46 ± 0.08	0.69995 ± 0.00015	20
Tieschitz	H3	4.53 ± 0.06	0.69880 ± 0.00010	11
Krähenberg	LL5	4.6 ± 0.3	0.6989 ± 0.0010	8
Olivenza	LL5	4.53 ± 0.16	0.6994 ± 0.0010	17
Saint Severin	LL6	4.51 ± 0.15	0.69903 ± 0.00020	9
Soko-Banja	LL4	4.45 ± 0.02	0.69959 ± 0.00024	12
Allende	CV3	~ 4.5	0.69877 ± 0.00002	7
Allende	CV3	~ 4.5	0.69880	21
Differentiated meteorites:				
Norton County	Aub	4.6 ± 0.1	0.700 ± 0.002	3
Norton County	Aub	4.39 ± 0.04	0.7005 ± 0.0004	10
Bereba	Euc	4.08 ± 0.26	0.69898 ± 0.00007	2
Ibitira	Euc	4.42 ± 0.25	0.69895 ± 0.00010	2
Juvinas	Euc	4.50 ± 0.07	0.69898 ± 0.00005	1
Sioux County	Euc	4.10 ± 0.14	0.69897 ± 0.00008	2
Stannern	Euc	3.2 ± 0.5	0.6993 ± 0.0012	2
Kapoeta†	How	3.5 – 3.8	0.69990	16
Kapoeta†	How	4.44 ± 0.12	0.69885	15
Colomera	IIE	4.51 ± 0.04	0.69940 ± 0.00004	18
Kodaikanal	IIE	3.7 ± 0.1	0.713 ± 0.020	4
Weekeroo Station	IIE	4.28 ± 0.13	0.703 ± 0.003	5
Nakhla	Acungr	1.34 ± 0.02	0.70232 ± 0.00006	14

° All ages are calculated for a ^{87}Rb decay constant of 1.42×10^{-11} yr^{-1}.

† Different fractions of the Kapoeta breccia yield different ages, the lower values presumably being metamorphic ages.

References:

1. C. J. Allegre, J. L. Birck, S. Fourcade, and M. Semet. 1975. *Science* 187:436.
2. J. L. Birck and C. J. Allegre. 1978. *Earth Planet. Sci. Lett.* 39:37.
3. D. D. Bogard, D. S. Burnett, P. Eberhardt, and G. J. Wasserburg. 1967. *Earth Planet. Sci. Lett.* 3:179.
4. D. S. Burnett and G. J. Wasserburg. 1967. *Earth Planet. Sci. Lett.* 2:137.
5. D. S. Burnett and G. J. Wasserburg. 1967. *Earth Planet. Sci. Lett.* 2:397.
6. K. Gopalan and G. W. Wetherill. 1970. *J. Geophys. Res.* 75:3457.
7. C. M. Gray, D. A. Papanastassiou, and G. J. Wasserburg. 1973. *Icarus* 20:213.
8. W. Kempe and O. Müller. 1969. In *Meteorite Research*, ed. P. M. Millman (Reidel), p. 418.
9. G. Manhes, J. F. Minster, and C. J. Allegre. 1978. *Earth Planet. Sci. Lett.* 39:14.
10. J. F. Minster and C. J. Allegre. 1976. *Earth Planet. Sci. Lett.* 32:191.
11. J. F. Minster and C. J. Allegre. 1979. *Earth Planet. Sci. Lett.* 42:333.
12. J. F. Minster and C. J. Allegre. 1981. *Earth Planet. Sci. Lett.* 56:89.
13. J. F. Minster, L. P. Ricard, and C. J. Allegre. 1979. *Earth Planet. Sci. Lett.* 44:420.
14. D. A. Papanastassiou and G. J. Wasserburg. 1974. *Geophys. Res. Lett.* 1:23.
15. D. A. Papanastassiou and G. J. Wasserburg. 1976. *Lunar Sci.* 7:665.
16. D. A. Papanastassiou, R. S. Rajan, J. C. Huneke, and G. J. Wasserburg, 1974. *Lunar Sci.* 5:583.
17. H. G. Sanz and G. J. Wasserburg. 1969. *Earth Planet. Sci. Lett.* 6:335.
18. H. G. Sanz, D. S. Burnett, and G. J. Wasserburg. 1970. *Geochim. Cosmochim. Acta* 34:1227.
19. G. J. Wasserburg and D. S. Burnett. 1969. In *Meteorite Research*, ed. P. M. Millman (Reidel), p. 467.
20. G. J. Wasserburg, D. A. Papanastassiou, and H. G. Sanz. 1969. *Earth Planet. Sci. Lett.* 7:33.
21. G. W. Wetherill, R. Mark, and C. Lee-Hu. 1973. *Science* 182:281.

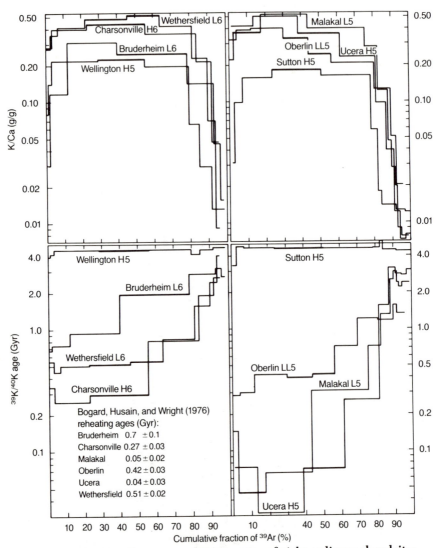

Figure III-3. The $^{40}K-^{40}Ar$ ages and K/Ca ratios of eight ordinary chondrites. The amounts of ^{39}K and ^{40}Ca are determined by neutron activation to ^{39}Ar and ^{37}Ar. Mineral fractions are sampled by outgassing argon from the sample by a series of heating steps. The 600°C–700°C fraction is indicated by the heavy bars (for two of the chondrites, two fractions each were collected in this range). Six meteorites showing textural and mineralogical evidence of heavy shock have low ages in the low-temperature fractions, gradually increasing to higher ages in the highest-temperature fractions. With minor exceptions, the two unshocked meteorites Wellington and Sutton have ages of 4.4 to 4.6 Gyr in all fractions. (Data from D. Bogard et al., *J. Geophys. Res.* 81:5664, 1976)

Figure III-3. Apparent ages and K/Ca ratios are calculated for each fraction. In a well-behaved system, the apparent ages at several adjacent temperature steps will be similar, resulting in a "plateau." If, as for the Sutton H5 and Wellington H5 chondrites, the plateau extends across all temperature steps, then the result is interpreted as a formation age ($\sim$ 4.5 Gyr in these cases). If the mineral fractions outgassing at lower temperatures yield lower ages than those outgassing at higher temperatures, then this is interpreted as evidence for a later metamorphic reheating that can be dated in favorable cases — i.e., when several of the lower-temperature fractions form a plateau. Thus, the Wethersfield L6 and Oberlin LL5 chondrites yield reheating ages of 510 $\pm$ 20 Myr and 420 $\pm$ 30 Myr, respectively. The remaining chondrites do not yield well defined plateaus, and their reheating ages have higher relative errors: Bruderheim, 700 $\pm$ 60 Myr; Charsonville, 270 $\pm$ 80 Myr; Malakal, 50 $\pm$ 20 Myr; and Ucera, 40 $\pm$ 30 Myr. These six meteorites having ages much less than 4.5 Gyr show strong shock effects, and the reheating ages are believed to date the times that major impacts occurred on the parent body. The ^{39}Ar released at low temperatures is associated with a phase having a high K/Ca ratio (perhaps a feldspar), whereas the phase(s) retaining ^{39}Ar to high temperatures have low K/Ca ratios (perhaps mainly clinopyroxenes). The phases having high K/Ca ratios apparently lost ^{40}Ar more rapidly in a brief reheating event.

Relative formation intervals based on the ^{129}I–^{129}Xe system ($T_{1/2}$ = 16 Myr) are also determined by a neutron-activation controlled-temperature release method similar to that used in the ^{39}Ar–^{40}Ar method. No ^{129}I remains in the sample but, if all minerals became gas-tight at the same time, the amount of stable ^{127}I in each mineral fraction is proportional to the amount of ^{129}I present when each mineral fraction became closed (i.e., when the retention of ^{129}Xe commenced). Capture of a neutron converts ^{127}I to ^{128}I, which then emits a beta particle (electron) to form ^{128}Xe. Mass-spectrometric analysis is used to determine, in each temperature (mineral) fraction, the amounts of ^{128}Xe resulting from neutron capture on I and of ^{129}Xe resulting from the decay of ^{129}I in excess over the amounts of ^{128}Xe and ^{129}Xe present in the primordial xenon (see discussion later in this chapter). From these data, one can calculate the $(^{129}I/^{127}I)_I$ value in each mineral fraction at the time xenon retention commenced; if these values show good agreement, then the mean ratio is a measure of the time of formation (a ratio two times greater indicates formation 16 Myr earlier).

Figure III-4 shows $(^{129}I/^{127}I)_I$ values determined in a number of meteoritic samples (chondritic samples, with the exception of the Shallowater and Peña Blanca Spring aubrites). The lower scale shows relative ages calculated from these ratios; zero is arbitrarily assigned to the position of the Bjurböle chondrite, which was used as a standard in the neutron irradiations.

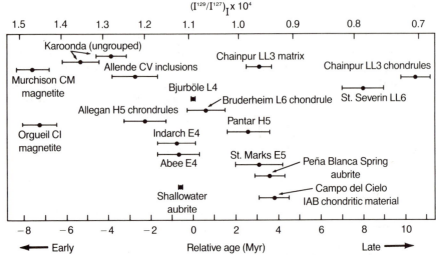

Figure III-4. Relative $^{129}I - ^{129}Xe$ formation intervals (Bjurböle $\equiv 0$), mainly determined by F. Podosek (1970). Note that the total range is only about one ^{127}I halflife (16 Myr), and that the differentiated aubrites began retaining xenon at about the same time as did the chondritic materials. Uncertainties may be somewhat greater than those indicated. (Data from F. Podosek, *Geochim. Cosmochim. Acta* 34:341, 1970)

The most important fact to emerge from these data is that all these meteorites formed within a 16-Myr interval. The significance of the difference between any two samples is obscured by the fact that the major iodine-bearing, xenon-retentive components have not been defined. If some of these components predate the agglomeration of the chondrites, then relative formation intervals may be subject to uncertainties significantly larger than those indicated by the bars, and the true range in formation times could be larger or (more likely) smaller than 16 Myr.

Formation interval information is also contained in the $(^{87}Sr/^{86}Sr)_I$ values determined from Rb–Sr mineral isochrons. If we assume that rubidium initially had a uniform isotopic composition, and that the nebular Rb/Sr ratio was uniform independent of distance from the Sun, then we can calculate relative formation intervals from precise $(^{87}Sr/^{86}Sr)_I$ values. A few examples are listed in Table III-2; data sources are restricted to a single research group to avoid interlaboratory calibration errors. Formation intervals are calculated from a rate of $(^{87}Sr/^{86}Sr)_I$ increase of 1.07×10^{-4} Myr^{-1}, based on the CI chondrite Rb/Sr ratio 0.25 ± 10. There is general agreement between these relative formation intervals and those based on the $^{129}I - ^{129}Xe$ system. The occasional discrepancies may in part indicate

Table III-2
Relative formation intervals based on measured $(^{87}Sr/^{86}Sr)_I$ values
and an assumed Rb/Sr atom ratio of 0.25

Meteorite	Class	$(^{87}Sr/^{86}Sr)_I$	Formation interval (Myr)
Allende	CV3	0.69877	$\equiv 0$
Angra dos Reis	Acungr	0.69884	7
Moore County	Euc	0.69891	13
Colomera	IIE	0.69940	~59
Peace River	L6	0.69970	~87
Guareña	H6	0.69995	110

Data from C. M. Gray et al., *Icarus* 20:213, 1973.

that iodine- and rubidium-bearing minerals did not become closed systems at the same time, and may in part be due to the problem mentioned earlier (that some mineral components may even predate the formation of the rock).

In some refractory-element–rich inclusions (mostly from the Allende CV chondrite), one finds ^{26}Mg excesses produced by the decay of extinct ^{26}Al $(T_{1/2} = 720$ kyr$)$. Because of the short halflife, these inclusions must have been cool enough to prevent isotopic exchange within a few megayears following the production of the ^{26}Al in a supernova explosion.

In contrast, the probable time interval between the formation of the ^{129}I in a supernova and the formation of chondritic rocks cool enough to retain ^{129}Xe is estimated to be about 100 Myr, based on simple models of supernova formation rates. There are various ways in which this discrepancy can be reconciled. One possibility is that the decay of ^{26}Al to ^{26}Mg predates solar-system formation — in other words, that the mineral grains in which this effect is preserved are genuine samples of interstellar grains. The other (more popular) explanation is that the ^{26}Al decayed after grains formed in the solar nebula. This implies that there were two supernova events predating solar-system formation, but that a negligible fraction of the ^{129}I in the early solar system was produced in the second event. This is not implausible because ^{26}Al is formed in an outer shell of a supernova, but ^{129}I is formed in a deeper shell or even in another class of star. If the $^{26}Al/^{27}Al$ ratio was as high in typical solar-system matter as the ratio measured in the Allende inclusions, ^{26}Al could have provided a major source of the heat needed to melt and differentiate parent bodies. The evidence discussed later in this chapter suggests that amounts of ^{26}Al were generally not sufficient to melt these bodies.

Cosmic-Ray Ages

Cosmic rays are high-energy particles and photons that are present in interplanetary and interstellar space. Most originate outside the solar system and are called **galactic cosmic rays.**

Galactic cosmic rays consist largely of protons having energies greater than 5 MeV, large enough to induce nuclear reactions. The mean energy is ~ 10 GeV, and the mean interaction depth for cosmic rays and the energetic secondary particles they engender is about 1 m. Thus meteoroids having dimensions of meters or less are subjected to continuous cosmic-ray bombardment in space.

Solar flares (violent storms in the solar atmosphere) sporadically produce large fluxes of low-energy particles; a small fraction of these particles have energies great enough to produce nuclear interactions in the outermost few millimeters of the meteoroid. These particles are sometimes called solar cosmic rays, but we will call them **solar-flare particles,** or solar-flare protons, and will restrict the term "cosmic rays" to extrasolar, galactic cosmic rays.

Cosmic-ray interactions produce both radioactive and stable nuclides. The relative amounts of radionuclides differing in halflife offer a measure of the variation in the cosmic-ray flux with time. The consensus interpretation of such data is that far from the Sun there has been no measurable variation in the flux of galactic cosmic rays during the past few gigayears. Within a few astronomical units (AU) of the Sun, variations in the low-energy end of the spectrum are produced by changes in the solar magnetic field.

Cosmic-ray exposure ages are the periods during which meteoroids have been irradiated by cosmic rays (i.e., during which the dimensions were a few meters or less). These are determined by measuring the amount of a stable **cosmogenic nuclide** and dividing this amount by the production rate. Many nuclides are produced by cosmic-ray interactions, but the cosmogenic fraction is generally not resolvable for nuclides of elements present in solar abundances. As a result, most studies are based on cosmogenic nuclides of elements having very low relative abundances. The nuclides of choice in chondritic meteorites are usually the least-abundant isotopes of the rare gases — for example, ^{21}Ne rather than ^{20}Ne or ^{22}Ne, and ^{38}Ar rather than ^{36}Ar or ^{40}Ar.

Because production rates vary with depth within a meteoroid, the determination of precise cosmic-ray ages requires measurement of both the production rate and the total accumulation of the stable product nuclides. As an example of such a technique, consider ages based on the simultaneous measurements of radioactive ^{81}Kr ($t_{1/2} = 210$ kyr) and stable ^{83}Kr; in chondritic meteorites, these nuclides are produced largely by nuclear interactions with strontium. Because the nuclides are very similar in mass number, the ratio of their production rates varies little with depth and, if (as commonly observed) the cosmic-ray age is greater than the ^{81}Kr half-life, very-precise cosmic-ray ages can be determined by this technique. Unfortunately, the low abundance of ^{81}Kr requires very-sensitive mass-spectrometric measurements, and only a few ^{81}Kr $-$ ^{83}Kr cosmic-ray ages have been reported.

Because of the difficulties associated with measuring production rates for each meteoritic sample, most published cosmic-ray ages are based on mean rates. In stony meteorites, the most-commonly-determined cosmogenic nuclides are ^{3}He and ^{21}Ne. In ordinary chondrites, the mean production rate of ^{3}He is about 2.0×10^{-14} cm$^3 \cdot$ g$^{-1} \cdot$ yr^{-1}; that for ^{21}Ne is about 3.5×10^{-15} cm$^3 \cdot$ g$^{-1} \cdot$ yr^{-1}; volumes are measured at 273 K and 1 atm. The ^{3}He production rate is nearly independent of depth, but the higher diffusion rate of the small ^{3}He atoms can lead to significant losses. The ^{21}Ne production rate has a moderate depth dependence, but diffusional loss is less severe than for ^{3}He.

Figure III-5 shows the distribution of ^{21}Ne cosmic-ray exposure ages in ordinary and carbonaceous chondrites. The data are not corrected for possible diffusional loss. Because of this problem and uncertainties in production rates, the precision of individual determinations is low (perhaps ± 30 percent of the tabulated value). Nevertheless, some trends are apparent. The H-group ordinary chondrites show a significant peak at about 5.5 Myr; it appears that a major breakup event produced H-group meteoroids at this time. There are no narrow peaks among L or LL chondrites. Regolithic breccias (those meteorites bearing solar-type rare gases) show essentially the same distributions as those not containing solar rare gases.

The carbonaceous chondrites also fail to show peaks; mean cosmic-ray ages in the groups show the trend CI < CM < CO $\cong$ CV. This ranking may partially be an artifact resulting from increasing gas loss with decreasing grain size from CV through CI, and it may partially be a real effect relating to decreasing survival periods in space resulting from increasing friability through the same sequence. Because there are few samples, the trend could conceivably reflect a fortuitous distribution of breakup events.

Cosmic-ray ages in a number of iron meteorites have been determined by a complex but precise method involving the measurement of the cosmogenic radioactive ^{40}K and stable ^{41}K, as well as rare gas nuclides used to estimate the depth below the original surface of the meteoroid. As illustrated in Figure III-6, cosmic-ray ages of irons are generally in the range 200 Myr to 1000 Myr. Two groups show a significant clustering of ages: with one exception, IIIAB ages are about 650 Myr; with two exceptions, IVA ages are about 400 Myr. Major disruptions of the parent bodies of these two groups must have occurred at these times.

It was earlier believed that the difference in cosmic-ray age between

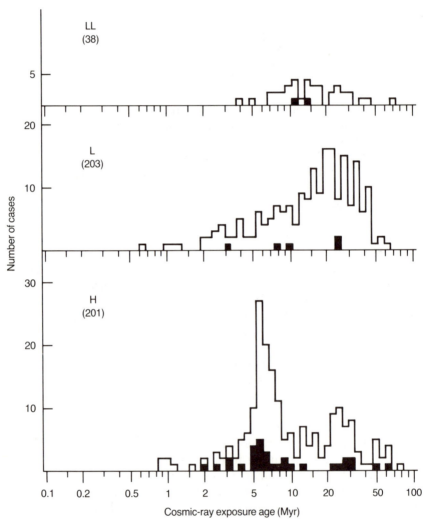

Figure III-5. Cosmic-ray age distribution in ordinary chondrites, based on ^{21}Ne contents. Most of the fluctuations appear to be random, but the large peak at 5.5 Myr in the H condrite distribution is too high to be random, and it indicates a major breakup event at that time. The shaded bars show the locations of chondrites containing solar rare gases, an indication of a regolithic history; their age distributions do not differ significantly from those of the remaining members of the groups. (After J. Crabb and L. Schultz, *Geochim. Cosmochim. Acta* 45:2151, 1981. Ages are systematically higher here because I have used a lower ^{21}Ne production rate.)

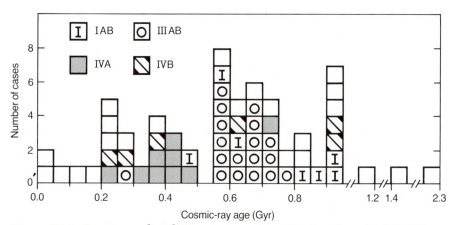

Figure III-6. Cosmic-ray distribution in iron meteorites based on the $^{40}K-^{41}K$ method. Most IIIAB irons fall in a cluster at 650 ± 100 Myr; most IVA irons fall in a cluster at 400 ± 100 Myr; the IAB and IVB ages scatter. The small number of ages less than 200 Myr partly results from the fact that meteorites with low ^{21}Ne ages were not studied. (Data from H. Voshage, Z. *Naturforsch.* 22a:477, 1967)

stones and irons reflected different spatial origins. Ages of ~10 Myr as observed in the stones are about equal to the mean lifetime of an object in an Earth-crossing orbit before either removal by a planetary collision or an orbital perturbation leading to ejection from the solar system. Similarly, a mean age of about 500 Myr for irons is comparable to the dynamic lifetime of a meteoroid in a Mars-crossing orbit.

A closer examination of possible sources shows that many stony meteoroids must be coming to Earth from Mars-crossing orbits or other asteroidal sources having long orbital lifetimes, and that the general absence of significant numbers of stony meteorites having ages greater than 50 Myr probably reflects their lower resistance to collisional destruction relative to the irons. This is a significant conclusion, for it shows that the flux of meteoroids entering the Earth's atmosphere is already biased in favor of the tougher irons. As noted in Chapter I, the rigors of atmospheric passage also add a bias in the same direction.

Terrestrial Ages

After a meteorite falls, it is shielded from cosmic rays by the Earth's atmosphere. Concentrations of cosmogenic radionuclides can be used to determine **terrestrial ages** for meteorite finds. For example, a typical chondrite having a cosmic-ray age greater than 3 Myr will have saturation levels of

270-yr ^{39}Ar, 5.7-kyr ^{14}C, and 300-kyr ^{36}Cl (where the time interval associated with each nuclide is its halflife) —i.e., the decay rates at the time of fall will be equal to the production rates in space. Each of these nuclides has been used to estimate terrestrial ages. If an ancient find is found to have a ^{14}C decay rate that is only 50 percent of the production rate, its terrestrial age is inferred to be 5.6 kyr. Occasionally terrestrial ages can be used to examine whether a newly found meteorite is associated with a historically recorded fall event. A recent example was a study of the Nogata (Japan) chondrite, which was found in a Shinto Shrine in a wooden box bearing the date May 19, 861 — 600 years earlier than the fall of the Ensisheim chondrite generally designated as the oldest observed fall (see Chapter I). To confirm the date, ^{14}C was measured in the wood from the box, and its date was determined to be A.D. 410 ± 350, even older than the age written on the box. It is hoped that ^{14}C terrestrial age can be determined directly on the chondrite in the near future.

An important recent application of terrestrial age dating resulted from studies of the meteorites from the Antarctic ice sheet. Several meteorites have terrestrial ages of ~700 kyr. These are the oldest terrestrial ages determined for stony meteorites, and they represent the first radiometric determination of the minimum age of the Antarctic ice sheet. Most previous estimates of the age of the ice had been ~100 kyr.

Outside Antarctica, the oldest measured terrestrial ages are 1.5 Myr for the Tamarugal IIIAB iron and ~100 kyr for the Potter L6 chondritic stone. Tamarugal is from the Atacama Desert of Chile, one of the driest regions in the world; because H_2O catalyzes the rusting process, weathering rates should be lower in dry climates. The heavily weathered Potter stone's resistance to disintegration resulted from its large size (more than 270 kg) and its burial in the sod of Nebraska in the central North American plains.

Solar-Flare Particles and the Solar Wind

I noted earlier that solar magnetic storms accelerate atoms of the solar atmosphere to relatively great energies — some as great as 100 MeV. Although these particles (mainly protons) produce nuclear interactions in the outer few millimeters of the meteoroid, most or all of the affected material is removed by ablation during passage through the Earth's atmosphere; as a result, detailed isotopic measurements of these effects have been confined to rocks exposed on the lunar surface and returned to Earth by the Apollo program.

The record of solar-flare particles is also recognizable as particle tracks. As discussed in Chapter II, many meteorites are breccias, and some of these show clear evidence of origins in planetary **regoliths,** the near surface layer mixed ("gardened") by repeated impacts. Irradiation of individual grains on the surface of the regolith by solar-flare protons produces damage to the crystal lattices. In favorable cases, this damage is not annealed out but is preserved in certain minerals of regolith breccia meteorites. Polishing and etching such grains converts these paths of damage to tracks that are readily visible under microscopic examination.

Figure III-7 shows an etched olivine crystal from a carbonaceous chondrite. The track density is heavy near the original surface but decreases rapidly in a direction perpendicular away from the surface. This depth dependence is that resulting from a typical spectrum of solar-flare particles. Because the solar-flare particles could not penetrate the nebular gas, this particle must have been irradiated in space or on the surface of a regolith after the uncondensed portion of the nebular gases had been swept out of the solar system.

Another property indicative of irradiation by solar particles is the pres-

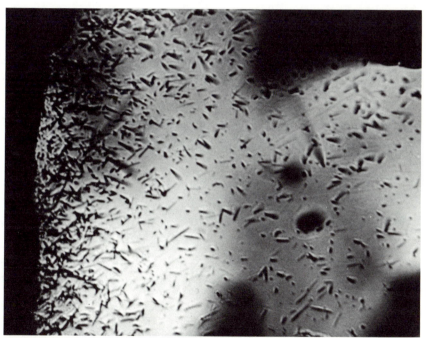

Figure III-7. Polishing and etching of this 150-μm olivine crystal from the matrix of the Murchison CM chondrite revealed many charged-particle tracks near the edges but few near the center. This track distribution is characteristic of irradiation by heavy (at least as massive as iron) solar-flare particles, most of which are not energetic enough to penetrate more than 20 μm into olivine. (Sample prepared and photographed by D. Macdougall)

ence of **solar rare gases.** The solid curve in Figure III-8 shows the abundance of major rare-gas isotopes in the solar atmosphere (see also Appendix D); this distribution is characterized by abundances decreasing exponentially with increasing mass number. The rare-gas distributions in some meteorites (such as Fayetteville and Pesyanoe in Figure III-8) are very similar to that in the solar atmosphere, the chief difference occurring at ^{132}Xe, which is two to ten times more abundant in the meteorites. Etching experiments show these gases to be concentrated near grain surfaces, as expected for the low energies (400 to 600 eV for each proton and neutron in the nucleus) of solar-wind ions. Similar rare-gas distributions are also found in lunar soils and breccias, and it is certain that irradiation by the solar wind is the dominant source of the rare gases having solar-type distribution. There is generally a strong correlation between the amounts of solar rare gases and of grains rich in solar-flare-particle tracks; with some exceptions, it appears

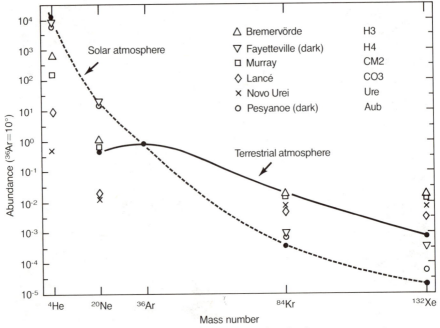

Figure III-8. The two types of patterns exhibited by the five rare gases in meteorites are shown for a major isotope of each element. Isotopic abundances are relative to ^{36}Ar abundance. The solar-type distribution is characteristic of samples from a planetary regolith; the planetary distribution is characteristic of chondritic and other primitive materials. (From J. Wasson, *Meteorites*, Springer, 1974)

that meteorites exhibiting these features originated in regoliths similar in many ways to the lunar regolith.

Major differences in the formation and evolution of these regoliths must have existed, however. Typical impact velocities on the Moon of 15 km · s^{-1} produce tough breccias showing substantial heating and metamorphism and other shock effects, sometimes including minor melting. The howardite, aubrite, ordinary-chondrite and enstatite-chondrite breccias are similarly tough, shocked rocks, and their shock effects seem consistent with similar or somewhat lower impact velocities. In contrast, CM and CI chondrites are relatively weak rocks that have not been reheated enough to release H_2O from their hydrated minerals; impact velocities in their "regoliths" must have been much lower, probably less than 1 km · s^{-1}.

The dashed curve in Figure III-8 shows the relative abundance of rare gases in the terrestrial atmosphere. No value for ^4He is shown because most **primordial** (present at the time of solar system formation) ^4He has escaped from the top of the atmosphere. The ^4He now in the atmosphere was produced mainly by the decay of ^{232}Th, ^{235}U, and ^{238}U and their decay products in the crust and mantle.

The relative abundances of the rare gases in most chondrites and in the differentiated ureilites are similar to those in the terrestrial atmosphere. For this reason, such distributions are designated **planetary rare gases;** they are marked by solar-normalized abundances that increase relative to solar abundances with increasing mass number (for example, Bremervörde, Murray, Lancé, and Novo Urei in Figure III-8).

The origin of the planetary rare-gas distribution is not yet known. It was once thought that the Earth's distribution reflects differing degrees of escape from the top of the atmosphere, but this model cannot hold for chondrites that formed directly in the solar nebula and show only minor amounts of alteration since their formation. Because the rare gases in the solar nebula surely had a solar distribution, it is likely that the planetary distribution was produced from the solar distribution by some process or combination of processes such as absorption on grain surfaces or diffusional escape from the interiors of mineral grains.

An important fact regarding planetary rare gases is that they are ubiquitous in chondritic (that is, undifferentiated) meteorites. Their presence can be resolved by careful mass-spectrometric measurements, even in the presence of solar-type gases in regolithic chondrites. For this reason, the presence of significant amounts of planetary-type rare gas is *prima facie* evidence that a meteorite without chondritic structure is subprimitive — that is, it was produced by minor alteration of chondritic material rather than by igneous differentiation. The "host" material in the Bencubbin breccia and the silicates in the IIE irons are examples of this class of subprimitive material.

Interstellar Grains and the Homogeneity of the Solar Nebula

The isotopic compositions of most elements are the same in all meteorite samples that have been studied. Among the studied elements are Cr, Ga, Mo, Cd, Sn, Ba, and Sm. The elements heavier than hydrogen are produced largely by elemental synthesis in stars. Because different kinds of stars produce different kinds of isotopes (proton-rich, slightly neutron-rich, or highly neutron-rich) and because the elemental matter in interstellar dust is expected to have originated in many different stars, the general absence of isotopic variations indicates that the solar nebula was initially well mixed. If the mean size of presolar solids was comparable to the millimeter size sampled for mass spectrometry, then good isotopic mixing seems to imply that preexisting solids were vaporized during the formation of the solar nebula, a conclusion in conflict with most current numerical models of solar-system formation by collapse of an interstellar cloud fragment (see Chapters VII and VIII).

Despite many searches and some false leads, there was no firm evidence for primordial isotopic anomalies until 1973; thus the solar system is indeed well mixed. The anomalies that have been found during the past decade offer exciting clues about the nature of presolar matter. However, with the exception of those in oxygen, these anomalies are small in magnitude and confined to a few kinds of meteorites — commonly to a few rare samples separated from those meteorites.

Three types of isotopic anomalies have been recognized in the components of certain chondrites: (1) unusual isotopic compositions in one or more rare-gas elements, attributed to unvaporized interstellar grains that constitute only a very minor fraction of the mass of the host chondrite; (2) unusual isotopic compositions in refractory elements that to date have been found in about 1 to 2 percent of investigated refractory inclusions from the Allende CV chondrite; (3) relative to terrestrial standards, ^{16}O-rich oxygen-isotope compositions found in all anhydrous materials (more than 50 percent) of chondrites belonging to the CM, CO, and CV carbonaceous chondrite groups, and ^{16}O-poor compositions found in chondrules from unequilibrated ordinary chondrites. The ^{16}O-rich compositions are also accompanied by titanium having an anomalous composition; in ^{16}O-rich samples, ^{50}Ti is very abundant relative to other titanium isotopes such as ^{44}Ti. A much smaller oxygen-isotope variation observed among the different groups of chondritic and differentiated meteorites probably also resulted from the process responsible for the more dramatic variations in CM, CO, and CV anhydrous minerals and chondrules.

In addition to the solar and planetary types of rare gases, both of which appear to have originated within the solar system, additional exotic kinds of rare gases are known. For example, a neon component having a very low $^{20}Ne/^{22}Ne$ ratio is probably contained in a few rare interstellar grains that survived infall into the nebula, and an unusual isotopic pattern in the heavy xenon isotopes is interpreted to be contained in another kind of interstellar particle. The latter anomaly seems to record the former presence of an unknown extinct nuclide that decayed by fission, perhaps an isotope of a superheavy element in the atomic-number range from 112 to 120. The excess of heavy xenon isotopes is always accompanied by an excess of light xenon isotopes that cannot be of fission origin, because fission always produces neutron-rich isotopes. Thus, even if the heavy isotopes are of fission origin, it appears that they are trapped in the same carrier grains with isotopes having a different source. If these carrier grains are of interstellar origin, there is no requirement that the fission of the parent have occurred after solar-system formation. It is at least equally plausible that the fissiogenic daughter products are fossil relicts of decays that occurred earlier.

Through 1982, researchers had studied magnesium from ~ 200 Allende refractory inclusions to investigate possible variations in the abundances of ^{26}Mg. In most inclusions the magnesium isotopic composition was normal. As discussed earlier, in a few inclusions, ^{26}Mg excesses correlate with the Al/Mg ratio and are thus attributable to in situ decay of ^{26}Al. In three inclusions, both ^{26}Mg and ^{25}Mg fractionations were found, the degree of fractionation of ^{26}Mg being twice that found for ^{25}Mg, as expected for mass-dependent fractionation (as we show later in this section). Several refractory elements (Ca, Ti, Ba, Nd) have also been found to show isotopic fractionation patterns that are different in each of these three inclusions. These fractionations typically involve only one or two of the six to eight isotopes of each element, and they are thus unrelated to the mass-dependent process that fractionated the magnesium isotopes. The source of these fractionations is unknown. These inclusions have been named FUN inclusions, the "F" referring to the fractionation effects observed in Mg, Si, and O, and "UN" to the unknown origin of the anomalous compositions of the refractory elements. These strange isotopic effects testify not only to the existence of unvaporized interstellar grains in the solar nebula, but also to a variety of isotopic compositions in the interstellar grains contributing to solar matter. Later in this section, these grains are discussed further in connection with their oxygen-istopic composition.

Because ^{26}Al has the potential to be a heat source for melting small asteroids (radius greater than 5 km), it is highly desirable to determine the mean $(^{26}Al/^{27}Al)_I$ ratio in solar matter at the time kilometer-sized bodies first formed. Several investigators have accepted the ratio of 5×10^{-5} found in numerous refractory inclusions in the Allende CV chondrite as this mean value. However, because this ratio is about eight times higher than that required to melt a chondritic body (see Chapter IV), it seems inconsistent with the large fraction of chondrites (meteorites never melted following accretion) and the rarity of igneously formed stones among meteorite falls.

A recent investigation that focused on Allende refractory inclusions that were melted late, possibly from impact during parent-body formation, has yielded $(^{26}Al/^{27}Al)_I$ ratios of 6×10^{-6} or less — at or below the level capable of melting a body of radius 5 km or greater. Final assessment of the efficacy of an ^{26}Al heat source must await further studies but, in my opinion, the weight of the current evidence indicates that ^{26}Al was not the chief heat source responsible for melting the parent bodies of the igneously formed stony and iron meteorites.

Figure III-9 illustrates the fractionation and mixing relationships among oxygen isotopes. The vertical axis represents $\delta^{17}O$, the difference in $^{17}O/^{16}O$ ratio between the sample and a standard, with the difference normalized to the standard composition; the horizontal axis represents $\delta^{18}O$, defined analogously.[1] The values are expressed per mil (‰) —that is, in parts per thousand. Note that, in effect, one is plotting a normalized $^{17}O/^{16}O$ ratio versus a normalized $^{18}O/^{16}O$ ratio.

Physical and chemical processes produce fractionations that depend on differences in mass. Samples produced by such fractionations in an initially homogenous system spread out along a line with a slope of $\frac{1}{2}$ (more exactly, 0.52) on a $\delta^{17}O - \delta^{18}O$ diagram; the slope chiefly reflects the ratio of the difference in mass between the minor isotope and ^{16}O. In Figure III-9, a line having a slope of 0.52 labeled terrestrial fractionation (TF) line shows the loci of points from a variety of terrestrial samples (rocks, natural waters, and so on). All terrestrial samples fall along this line.

In 1973, R. N. Clayton and colleagues discovered that the oxygen-isotopic composition in carbonaceous-chondrite anhydrous minerals (CCAM) describes a line with a slope of ~1; an updated version of the line (the slope is now determined to be 0.94 ± 0.01) appears in Figure III-9. The line cannot have been generated by fractionation but must result from mixing material from different reservoirs, one at least as ^{16}O-rich as the lower extreme of the line and, similarly, one at least as ^{16}O-poor as the upper extreme. In fact, if there were *only* two solar-system reservoirs, the extrapolated composition of the ^{16}O-poor component must be at least 2‰ richer in ^{17}O than the terrestrial fractionation line in order to account for the highest measurements reported in chondrules separated from L and LL chondrites.

---

[1] The equation for calculating $\delta^{17}O$ is the following:

$$\delta^{17}O(‰) = \frac{(^{17}O/^{16}O)_{\text{sample}} - (^{17}O/^{16}O)_{\text{standard}}}{(^{17}O/^{16}O)_{\text{standard}}} \times 1000$$

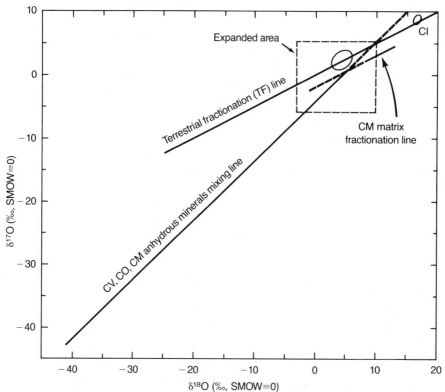

Figure III-9. The vertical axis of this diagram shows the difference between the chondrite $^{17}O/^{16}O$ ratio and that of the SMOW (standard mean ocean water) standard, relative to the measured composition of the standard and multiplied by 1000; the horizontal axis shows $^{18}O/^{16}O$ data treated similarly. All terrestrial samples lie along the terrestrial fractionation (TF) line. The CI chondrite and the other three carbonaceous chondrite groups show wide variations in oxygen-isotope composition. Most other meteorites lie inside the ellipse in the "expanded area," which is shown in more detail in Figures III-10, V-5, and IX-9.

Details regarding meteorites, meteorite groups, and meteorite components having bulk compositions inside the dashed rectangle are given in Figures III-10, V-5, and IX-8 and are discussed in connection with those diagrams. Bulk oxygen-isotope data for most chondrites and differentiated meteorites fall within an oval region straddling the terrestrial fractionation line in the $\delta^{18}O$ range 3‰ to 6‰. This observation plus the fact that the Earth, the Moon, and the CI chondrites lie near the terrestrial fractionation line suggest that the ^{16}O-poor reservoir (or reservoirs) contributed much more material to the solar nebula than did the ^{16}O-rich reservoir. The CM,

CO, and CV chondrites have bulk oxygen-isotope compositions well below the terrestrial fractionation line; even the fine-grained, iron-rich, low-temperature (see Chapter VIII) fraction of these materials plots below the terrestrial line. This observation indicates that significant fractions of the elements now in the anhydrous minerals of these chondrites were not vaporized during infall into the solar nebula and either (1) that nebular gas at the CM, CO, and CV formation locations did not have terrestrial oxygen-isotopic composition falling on the terrestrial fractionation line, or (2) that none of the investigated samples (not even the fine matrix) completely equilibrated with the gas.

In Figure III-10, a portion of Figure III-9 is expanded to show data on chondrules separated from unequilibrated ordinary chondrites (OCC) and data on matrix, chondrules, and some anhydrous mineral samples from carbonaceous chondrites. Most of the anhydrous-mineral samples are from the Allende CV chondrite, and we have plotted only Allende data in Figure III-10. Also shown in Figure III-10 are the fields occupied by the Earth's mantle, H, L, and LL chondrites, and the basaltic eucrites. As noted earlier, the Allende samples yield a well-defined array with slope 0.94 ± 0.01. Interestingly, the chondrule samples yield a roughly similar slope (0.92 ± 0.23), though with a much larger uncertainty. However, the line segments intersect the terrestrial fractionation line at points differing by about 6‰ to 7‰ in $\delta^{18}O$.

The chief arguments for believing that the nebula lay along the TF line are that the Earth and the Moon formed near the Sun, where nebula temperatures were probably high enough to vaporize most presolar solids and yield extensive isotopic exchange between any unvaporized solids and the gas. Another clue is that the CI chondrites that consist almost entirely of fine-grained (less than 1 μm) solids have oxygen-isotope compositions very near the terrestrial fractionation line. It seems probable that they either formed by condensation from or underwent extensive isotopic exchange with the gas. Thus, their composition probably indicates a mean gas composition near the terrestrial fractionation line (Figure III-9). The solar abundance of oxygen is about 19 times those of the major oxide-forming elements magnesium and silicon; thus one can show that, at temperatures higher than 1000 K, about 17 percent of the total oxygen was tied up as MgO, SiO_2, and minor oxides—the nebular gas containing the remaining ~83 percent. The gaseous fraction was larger at temperatures high enough to evaporate an appreciable fraction of the presolar oxides.

The high slopes of the CCAM and OCC arrays imply that they were formed by the mixing of components from distinct reservoirs—that is, reservoirs formed by different nuclear or other processes and thus never in equilibrium. Although some photochemical processes can produce arrays with a slope of unity, it is difficult to envision how such processes could have

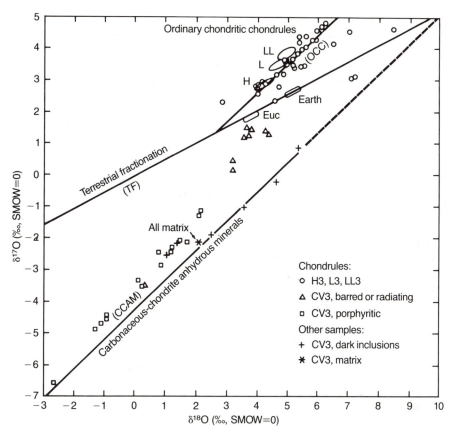

Figure III-10. On a $\delta^{17}O - \delta^{18}O$ diagram, two sets of meteoritic samples form arrays with slopes much steeper than 0.52, the value resulting from fractionation. These arrays must have resulted from the mixing of isotopically distinct materials. A plausible suggestion is that the most-^{16}O-rich materials (near $\delta^{17}O = -42\%_o$, $\delta^{18}O = -40\%_o$) in the carbonaceous-chondrite-anhydrous-mineral (CCAM) array and the most-^{16}O-poor materials (near $\delta^{17}O = +5.0\%_o$, $\delta^{18}O = +6.5\%_o$) in the ordinary-chondrite-chondrule (OCC) array represent independent reservoirs of presolar materials; the arrays formed by reaction of these materials with nebular gas have a composition near the terrestrial fractionation (TF) line. For reference, the positions of the Earth, the basaltic eucrites (Euc), and the H, L, and LL chondrites are shown. (Data mainly from R. Clayton et al. in *Chondrules and Their Origins*, ed. E. L. King, Lunar Planet. Inst., 1983, p. 37)

played a major role in a solar nebula that was probably opaque throughout the period before planetesimal formation. Also, photochemical reactions lead to minor reservoirs that are ^{16}O-poor, whereas (as discussed earlier) the minor solar-system reservoir was ^{16}O-rich. Because the two slope-one lines on Figure III-10 are not congruent, it is clear that the same two components cannot account for both arrays, and that at least one more component is required.

A plausible idea is that the most-^{16}O-rich CCAM component and the most-^{16}O-poor OCC component were each presolar solids, and that the intermediate component was the nebular gas. The fact that the upper end of the CCAM line and the lower end of the OCC line are offset can be explained by the fact that the isotopic composition of solids formed by reaction with the gas should lie at different positions along the terrestrial fractionation line, depending on the mineral phase formed and the temperature. The offset towards lower $\delta^{18}O$ values of the OCC–TF intersection from that of the CCAM–TF intersection is in the direction expected if reaction temperatures in the region where ordinary chondrites formed were higher than those producing the ^{16}O-poor CCAM component. According to such a picture, the mean nebular gas composition may have been about 6‰ to 8‰ $\delta^{18}O$ on the TF line.

The picture just described is only a very simple cartoon of a highly complex nebular picture. Many will find it unsatisfying, but a more detailed picture is neither consistent with the goals of this book nor warranted by my confidence in the accuracy of the depiction I can currently achieve. For present purposes, the important facts are that both the ordinary chondrites and the CM, CO, and CV carbonaceous chondrites preserve different isotopic signatures of presolar matter, but that all samples plotted in the two arrays have experienced severe physical alteration in nebular and, to some degree, in parent-body processes.

A few inclusions exist in which the preservation of unique isotopic anomalies in several elements indicates that alteration has been minimal. In three inclusions from CV Allende, the oxygen isotopes in separated minerals do not fall along the anhydrous-minerals mixing line (slope = 0.94) but form their own arrays with differing slopes. As shown in Figure III-11, these arrays are on the high-$\delta^{18}O$ side of the CCAM line, and they and the CCAM line tend to converge near $\delta^{17}O = 0$, $\delta^{18}O = 4‰$. Interestingly, a line through the mean composition ($\delta^{17}O = -42‰$, $\delta^{18}O = -40‰$) of the most ^{16}O-rich normal inclusions and the spinel fraction of the EK 1-4-1 inclusion has a slope of 0.52; the more-pure hibonite sample from the HAL inclusion falls near this line. Apparently the aluminum-rich phases spinel ($MgAl_2O_4$) and hibonite ($CaAl_{12}O_{19}$) were most resistant to alteration by reaction with nebular gas. Their positions along the same line of slope 0.52 suggest that they formed by fractionation of a single starting material,

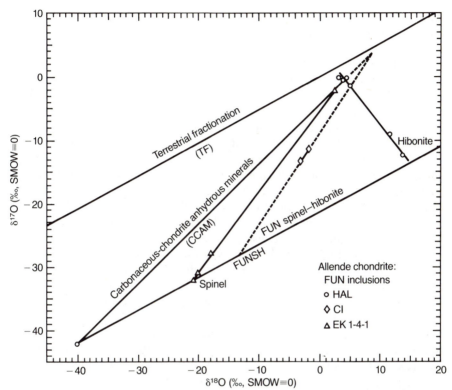

Figure III-11. A few rare inclusions in the Allende CV chondrite show mass fraction (F) effects in Mg and anomalous isotopic compositions of unknown nuclear (UN) origin in refractory elements such as calcium and titanium. On an O-isotope diagram, these inclusions fall in an otherwise uninhabited region on the high-^{18}O side of the CCAM line. The oxygen-isotope compositions of the most-aluminum-rich phases $MgAl_2O_4$ and $CaAl_{12}O_{19}$ in two of these inclusions fall near a fractionation line passed through the most-^{16}O-rich of the "normal" Allende inclusions near $\delta^{17}O = -42\%_0$, $\delta^{18}O = -40\%_0$. These FUN inclusions attest to the isotopic variety of the presolar materials that came together to form the solar nebula. (Data from T. Lee et al., Geophys. Res. Lett. 7:493, 1980, and earlier work cited therein)

perhaps by distillation of presolar material having $\delta^{17}O = -42\%_0$, $\delta^{18}O = -40\%_0$.

As noted earlier, each of these three inclusions exhibits an isotopic fractionation of magnesium; a plot of $^{25}Mg/^{24}Mg$ ratio against $^{26}Mg/^{24}Mg$ ratio gives a slope of 0.5. The refractory elements such as Ca, Ti, and Ba, however, show anomalies that are unique in each inclusion; because these anomalies are often localized in only one or two out of five to eight stable isotopes, they cannot be due to fractionation. For this reason they are assigned an "unknown nuclear" origin. As noted earlier, these three inclusions are called FUN inclusions (F for *f*ractionation, UN for *un*known nuclear). Clearly, the isotopic alteration of the aluminum-rich phases of these inclusions in the solar nebula was minimal, and their precursor components were produced under differing stellar (and possibly interstellar) conditions. They are very rare; only three of the first 200 investigated igneous-textured inclusions show the FUN set of anomalies. None show any ^{26}Mg excess resulting from ^{26}Al decay; if the ^{26}Mg excesses in other inclusions are the "fossil" record of ^{26}Al decay in grains that predate the solar system, then one might expect to find such excesses in the FUN inclusions. Their absence could mean (1) that the elements showing anomalous isotopic compositions were not formed in the supernova responsible for the ^{26}Al, or (2) that the suggested distillation process occurred at a later time, after ^{26}Al had decayed; other explanations are also possible.

In the matrix of CM chondrites, oxygen-isotopes compositions form an array with a slope of roughly 0.5 that has often been interpreted to be the result of chemical or physical mass-dependent fractionation processes. If this slope is due to fractionation, then either (1) this fractionation occurred in a parent body that had an oxygen-isotopic composition different from that of nebular gases, or (2) the nebular gases did not have the same oxygen-isotope composition at the CI and CM locations. In fact, the slope of the matrix array is somewhat greater than 0.5, and it appears that mixing played too large a role to allow strong conclusions about fractionation processes.

Suggested Reading

Begemann, F. 1980. Isotopic anomalies in meteorites. *Rep. Prog. Phys.* 43:1309. A technical review of isotopic anomalies in meteorites, with brief discussion of their origins and other implications.

Faure, G. 1977. *Principles of Isotope Geology.* Wiley. Detailed discussions of radiometric dating methods and of isotopic fractionation.

Kirsten, T. 1978. Time and the solar system. In *The Origin of the Solar System*, ed. S. F. Dermott, (Wiley), p. 267. A technical review of radiometric and cosmic-ray ages of meteorites, and a synthesis of early solar-system chronology.

Podosek, F. A. 1978. Isotopic structures in solar system materials. *Ann. Rev. Astron. Astrophys.* 16:293. A technical review of the isotopic variations observed in meteoritic samples with the focus on nonradiogenic effects in the rare-gas elements.

Schramm, D. N., and R. N. Clayton. 1978. Did a supernova trigger the formation of the solar system? *Sci. Amer.* 239(4):124. Discussion of the meteoritic isotopic anomalies associated with interstellar grains, and the possibility that a supernova explosion triggered the formation of the solar system.

Wasson, J. T. 1974. *Meteorites: Classification and Properties.* Springer. Chapters 8 and 9 cover variations in stable isotopes, Chapter 10 radio-metric ages, and Chapter 11 cosmic-ray ages.

Wetherill, G. W. 1975. Radiometric chronology of the early solar system. *Ann. Rev. Nucl. Sci.* **25**:283. A technical review of formation intervals, formation ages, and metamorphism ages, with a discussion of how these relate to events occurring early in solar-system history.

CHAPTER IV

Iron Meteorites: Evidence For and Against Core Origins

There is no doubt that some meteorites were formed by melting in planetary settings. The basaltic meteorites are compositionally closely related to the terrestrial and lunar basalts. **Basalts** are **extrusive rocks** originally formed by melting in the interiors of these planets, then transported to the cold surface ("extruded") where they rapidly solidified. The formation of these differentiated silicate-rich meteorites is discussed in Chapter V. A key fact is that the common meteoritic basalts had already been melted and extruded ~ 4.4 Gyr ago, within ~ 100 Myr of the birth of the solar system.

For many years, it was assumed that iron meteorites represent the metallic cores of planets that had experienced extensive degrees of melting and phase separation. We now know that this model is not correct for all iron meteorites; for most groups, however, the detailed evidence now available (and summarized later in this chapter) demonstrates that these iron meteorites are indeed fragments of the cores of small planets.

A key (and, as we will see, unanswered) question is, What was the heat source that melted and differentiated these parent bodies? The smaller the body, the stronger a heat source (expressed in the rate of heat production per unit mass) must be, primarily because heat is generated at a rate proportional to the volume of the body but is lost from the surface, and the surface-to-volume ratio increases with decreasing radius of the body. Thus the best way to approach the formation of the differentiated meteorites is to survey the possible heat sources.

Planetary Heat Sources

The planetesimals that accreted to form the parent bodies of the differentiated meteorites probably either were chondritic or had similar compositions differing from those of known chondrites only in the degree of fractionation (refractory-element or volatile-element fractionations) produced by nebular processes. In order to differentiate the silicates and separate the metal into cores, the viscosities of the materials in these bodies had to be small enough to allow the immiscible metal and silicate phases to undergo gravitational separation, with the denser ($\rho = 7 \pm 1 \text{ g} \cdot \text{cm}^{-3}$) mixture of Fe–Ni and FeS segregating to the center of the parent body, and the less-dense ($\rho \cong 3.3 \pm 0.3 \text{ g} \cdot \text{cm}^{-3}$) silicates forming a mantle.

The rate at which metal settles into the core depends on the viscosity of the dominant silicates. At about 1600 to 1650 K, chondritic material is largely molten, and the viscosity becomes about the same as that of molasses at room temperature, small enough to allow core formation to occur. We don't yet have a means of dating the time of formation of the cores parental to the iron meteorites but, because of the short-lived nature of the most-plausible heat sources, it surely occurred within 100 Myr of the time of parent-body formation.

Let us assume that the initial temperature throughout the parent body was about 250 K — higher than the present temperature in the asteroid belt, but lower than the inferred nebular temperature (400 to 500 K) when condensation ceased (see Chapter VII). Then, for each gram of material, about 1000 J were needed to bring the temperature to 1600 K, and about 300 to 400 J more were needed to melt it. Thus, to obtain melting, the minimum heat input required was ~ 1300 J $\cdot$ g^{-1}. Additional heat was required if some was lost by leakage out of the surface. If hotter than the surroundings, the surface loses heat by radiation in the infrared.

Possible sources of the heat that melted the parent bodies of the differentiated meteorites include (1) the long-lived radionuclides such as ^{40}K, (2) extinct radionuclides such as 720-kyr ^{26}Al, (3) radiation from a superluminous phase of the Sun, (4) electric currents induced by immersion of the planet in a superintense solar wind, and/or (5) conversion of gravitational or kinetic energy to heat during accretion.

Chondritic abundances of the long-lived radionuclides ^{40}K, ^{232}Th, ^{235}U, and ^{238}U provide enough heat to produce melting in an Earth-sized (radius = 6380 km) planet but, because of the higher rate of heat loss from the surface, are not sufficient to melt a Moon-sized (radius = 1740 km) object, even when their abundances are increased to correct for decay during the past 4.5 Gyr. As will be discussed in more detail later, the radii of the meteorite parent bodies were no larger than about 500 km, and most were probably considerably smaller. Unless the long-lived radionuclides were in concentrations far in excess of chondritic values, they were not a major source of the heat that melted some meteorite parent bodies.

A chondritic parent body with radius greater than 50 km will melt if its ^{26}Al concentration is greater than 100 ng/g. If the ^{27}Al concentration is ~ 12 mg/g, as in ordinary chondrites, then melting would have occurred if the initial ^{26}Al/^{27}Al atom ratio was greater than 8×10^{-6}. As discussed in Chapter III, ^{26}Mg excesses produced by ^{26}Al decay have been observed in a few refractory inclusions, most of them separated from the Allende CV chondrite. In most of these samples, the initial $(^{27}\text{Al}/^{27}\text{Al})_I$ ratio was $\sim 5 \times 10^{-5}$, enough to melt the interior of a body having a radius as small as 5 km. In contrast, most refractory inclusions have initial ^{26}Al/^{27}Al ratios of 2×10^{-6} or less. The upper limit also applies to two ordinary chondrites, but the investigated meteorites (L6 Bruderheim and H6 Guareña) show evidence of metamorphic recrystallization. If this recrystallization occurred several Myr after rock formation, then it could have equilibrated the magnesium isotopes and destroyed an earlier record of ^{26}Al. Because of the rarity of samples bearing a record of high $(^{26}Al/^{27}Al)_I$ ratios and the possibility that some of these are interstellar grain fossils, there is no way to determine the mean $^{26}Al/^{27}Al$ ratio in the nebula. As stated in Chapter III, I interpret the available data to indicate that the average ratio was probably too low to serve as the heat source that melted the differentiated-meteorite parent bodies, but this conclusion is by no means firm.

The gravitational collapse of matter to form the Sun converted large amounts of potential energy to heat. If the heat was transported to the surface by convection, then the luminosity of the Sun would have been many times (perhaps 500 times) higher than it is at present. If the Sun's mass were 1.5 times larger, the planets' orbital radii would have been 0.67 times the present values. Under these circumstances, the surface temperature of a blackbody (perfectly absorbing and emitting object) at the distance equivalent to present-day 1 AU could have reached 1620 K, and that at 2.8 AU about 970 K. Thus, if the Sun underwent such a highly luminous phase, then the surface layers of objects at 1 AU could have melted and differentiated, but those in the asteroidal region would have reached only metamorphic temperatures. Theoretical calculations indicate that the hypothetical highly luminous phase would have lasted only about 10^3 yr. In this period, heat could have diffused to a depth of about 200 m, but mixing connected with differentiation and impacts might have led to somewhat-greater heat penetration. As will be discussed in more detail later in this chapter, typical formation depths of iron meteorites appear to have been tens of kilometers or greater. It therefore seems doubtful that a superluminous Sun is an adequate heat source. An *ad hoc* and somewhat implausible scenario would involve heating of the material while in the form of 100 to 500-m planetesimals, and then accumulation of these while hot to form parent bodies having radii of tens or hundreds of kilometers.

Doppler-broadened absorption lines in the envelopes around T-Tauri stars (named after the type star in the constellation Taurus; see Chapter VI) imply that these stars are blowing off their atmospheres at rates about 10^7 times greater than the rate at which our Sun is losing matter through the solar wind. Such a T-Tauri wind would, like our solar wind, be a plasma (a partially ionized gas). Passage of such a plasma around an asteroid-sized body can introduce electrical currents that heat the body. Recent calculations by C. P. Sonett and colleagues indicate that this mechanism is most effective for bodies having radii in the range from 50 to 100 km, but that appreciable heating also occurs in bodies in the range from 25 to 50 km and from 100 to 250 km. The maximum temperatures reached at 2.8 AU are estimated to be about 1200 K, too low to differentiate a parent body, but temperatures near the 1600 K needed for melting and differentiation are realized near 1 AU.

Although this heat source looks promising, there are great uncertainties regarding the rate at which the wind's intensity decreased with time, the effect of rotation of the parent bodies, the effect of the large amounts of dust still present in the nebular midplane during these early stages of accretion, and even whether the Sun ever generated a T-Tauri wind. At best, this mechanism could lead to differentiation of parent bodies in the inner solar system, but the probability seems fairly low that it produced temperatures high enough to melt the parent bodies of the differentiated meteorites.

Two kinds of energy are released during planetary accretion: kinetic and gravitational. Kinetic energy results from the differences in relative velocities between the accreting mass and the planet, differences that are independent of the gravitational attraction between these bodies. Gravitational energy is the energy released due to mutual gravitational attraction of materials having no difference in relative velocity. Kinetic energy of accretion can release large amounts of energy on the surface of a planet. For example, extensive melting of the Earth's surface can sometimes be associated with the impacts of large meteoroids — such as that at Manicouagan, Quebec, Canada.

Impact heating on asteroid-sized bodies is less effective, however, because much of the strongly heated material is ejected from the crater with velocities exceeding the **escape velocity** — the vertical velocity that is just sufficient to remove matter from the surface of a body to an infinite distance (or, more practically, to a point where the gravitational attraction of the original body is no greater than that of other large bodies). The escape velocity v_{esc} (in $m \cdot s^{-1}$) from bodies having the density of ordinary chondrites ($\sim 3.5 \ g \cdot cm^{-3}$) is given by

$$v_{esc} \cong 1.4R \qquad \qquad \text{(IV-1)}$$

where R is the planetary radius in kilometers. Thus, the escape velocity from a 100-km asteroid is about $140 \ m \cdot s^{-1}$. If an object impacts a larger body at a velocity several times greater than the body's escape velocity, the amount of ejecta having velocities exceeding the escape velocity will be greater than the mass of the projectile, and the parent body will lose mass as a result of the impact.

It is clear from this type of argument that, during the period of parent-body growth, the relative accretion velocity cannot have been much greater than the escape velocity. If all kinetic energy is converted to heat and the heat confined entirely to the projectile, then the velocity needed to supply the $1300 \ J \cdot g^{-1}$ required to melt all accreting matter is $1.6 \ km \cdot s^{-1}$, about ten times the escape velocity of an asteroid of 100-km radius. Because some kinetic energy was not converted to heat, and some heat was lost by radiation from the surface between impact events, we conclude that impact heating did not produce large magma bodies in asteroids having radii less than ~100 km during the stage when these bodies were growing. Large impacts at a later (destructive) stage could have produced melting, but this melting would generally have been confined to the margins of the impact-generated crater, and experience with terrestrial and lunar impact melts shows that these melts usually solidify rapidly without producing appreciable amounts of igneous differentiation.

The mean amount of gravitational heat Q (in $J \cdot g^{-1}$) released in planet formation is given by

$$Q \cong (1.7 \times 10^{-4}) f \rho R^2 \qquad \text{(IV-2)}$$

where f is the fraction of the potential energy retained as heat, ρ is the density (in $g \cdot cm^{-3}$), and R is the radius (in km). Thus, even if f were unity (that is, no energy were lost by radiation to space), the total amount of gravitational energy released during formation of a body as large as the largest asteroid, Ceres ($R = 500$ km, $\rho \cong 3$ $g \cdot cm^{-3}$), is only ~130 $J \cdot g^{-1}$, a factor of 10 smaller than the minimum required for melting.

In summary, the heat source that melted the parent bodies of the differentiated meteorites remains uncertain. The most-probable sources are heating by extinct [26]Al and by currents induced by a T-Tauri-like solar plasma wind, but there are reasons to doubt that either of these sources was adequate for the task.

Elemental Fractionations Among Iron Meteorites

Figure II-10 shows large fractionations of gallium and germanium among all iron meteorites, but small fractionations in groups other than IAB and IIICD. Figure II-11 illustrates the large iridium fractionations within individual groups, the iridium range reaching factors of 6000 and 2000 in groups IIAB and IIIAB, respectively.

The groups having small ranges of germanium concentration form similar patterns on log–log plots of elements X versus nickel. Figure IV-1 illustrates data from three large groups (IIAB, IIIAB, and IVA) having small germanium ranges and from IAB, a group having a large germanium range. The general similarity in the patterns of the IIAB, IIIAB, and IVA groups implies that they formed by the same process. The simplest and most probable process appears to be fractional crystallization; there is general agreement that this is the only plausible process that could account for the fractionation of iridium by factors of several thousand. We will designate as **magmatic iron meteorites** those groups showing the patterns associated with formation by fractional crystallization, and as **nonmagmatic iron meteorites** the other two groups, IAB and IIICD.

If a molten body slowly solidifies under conditions such that there is no mixing (that is, by diffusion) in the solid and complete mixing in the liquid, then the Rayleigh equation

$$X_s = k_X X(1 - g)^{k_X - 1} \qquad \text{(IV-3)}$$

applies, where X_s is the concentration of element X in the solid when a fraction g of the original mass having a mean concentration X has solidified; k_X is the solid/liquid distribution ratio—i.e., the concentration of X in the solid divided by that in the liquid. One can show that a plot of $\log X$ versus $\log Ni$ should yield the linear relationship

$$\log X = A \log Ni + B \qquad \text{(IV-4)}$$

where $A = (k_X - 1)/(k_{Ni} - 1)$ provided that k_X and k_{Ni} remain constant during the crystallization of the core. From diagrams such as Figure IV-1, it is found that many elements do form linear arrays, implying distribution ratios that are nearly constant.

Experimental determinations of nickel distribution ratios in pure Fe – Ni generally fall in the range 0.7 to 0.9. Data from various sources suggest the use of a k_{Ni} vlue of ~ 0.85 for modeling the formation of most of the magmatically differentiated iron meteorites. Choosing this value, we can calculate k_X values for the other trace and minor elements ranging from ~ 4 for iridium to ~ 0.1 for phosphorus. Note that, because k_{Ni} is less than unity, all other elements having $k_X < 1$ are positively correlated with nickel, whereas elements having $k_X > 1$ show negative correlations with nickel. There is good qualitative agreement between k values inferred from $(\log X) - (\log Ni)$ plots and values determined in laboratory studies, although the meteoritic values tend to be more extreme — that is, farther from unity — as expected if equilibrium were more-nearly approached in meteorite formation than in the laboratory experiments.

The chief arguments in favor of a fractional-crystallization origin of the magmatic groups of iron meteorites are (1) the qualitative agreement between k_X values inferred from fractional-crystallization interpretation of iron-meteorite data and those determined by laboratory studies, (2) the large fractionations of refractory siderophiles (by factors as large as 6000)

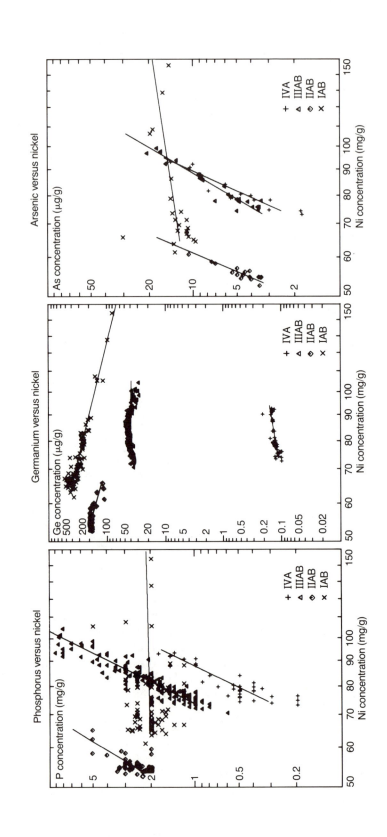

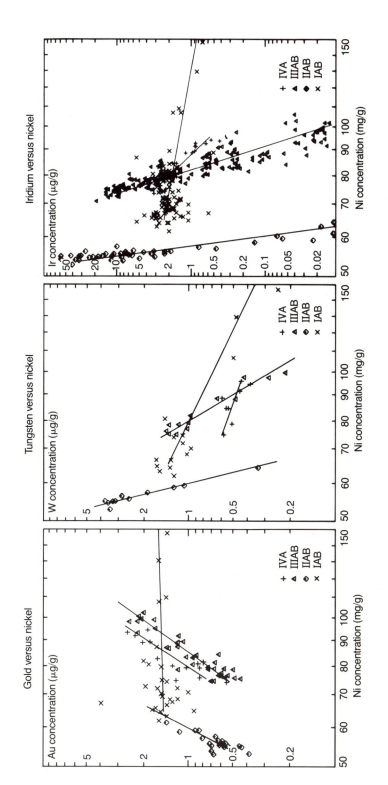

Figure IV-1. Diagrams showing the element/Ni distribution of P, Ge, As, Au, W, and Ir in the four largest groups of iron meteorites. The magmatic groups IIAB, IIIAB, and IVA probably formed by fractional crystallization of cores; the process(es) responsible for the fractionation of IAB are poorly defined. The positive correlations of P, As, and Au with Ni in the magmatic groups indicate solid/liquid distribution ratios less than unity; the negative correlations of W and Ir with Ni indicate ratios greater than unity; the germanium ratio is approximately unity.

that seem impossible to generate by any other mechanism, and (3) the general similarity in the patterns in all groups, implying that the same process can account for all.

The distinctly different patterns observed in the groups IAB (Figure IV-1) and IIICD implies that these groups have not formed by the same processes as the magmatic groups. These two groups contain silicates having chondritic compositions, and they have (mainly in the silicates) high planetary-rare-gas contents; these features are not found in the magmatic groups. The presence of silicates indicates that these nonmagmatic groups cannot have been completely molten: the silicates would have risen to the top of the metallic melt in a geologically short time. Furthermore, at temperatures high enough to melt the metal, the silicates would have differentiated — that is, would no longer be chondritic — and the rare gases would have been lost. On the basis of these arguments, it is clear that these groups were never fully molten — in other words, they were never part of a large metal–FeS magma. Rather, the metal and silicates appear to be mechanical mixtures that were incompletely melted (perhaps by shock) and then rapidly cooled to below the melting temperature.

In most IAB and IIICD irons, the amounts of silicates are minor (0.1 to 1 percent by mass). If melting followed by gravitational separation of immiscible metal and silicate did not occur, then how were the high concentrations of Fe–Ni generated from a starting chondritic mixture that included more silicates than metal? The most plausible idea appears to be mobilization and separation of shock-produced melts produced by impacts in the near-surface region of the parent body. The composition of the resulting metal may have depended on the temperature reached by the shock melt: the low temperature melts had high nickel contents, and the high-temperature melts have low nickel contents. The higher the temperature, the more efficient the extraction of germanium, iron, and iridium out of nebular oxide phases into the metal–FeS melt. Other **scenarios** also have been proposed for the formation of these puzzling meteorites. (I use **model** for a relatively well-defined set of conditions and processes that are proposed to explain a set of observations, but **scenario** for a more-sketchy set of conditions and processes that involves more assumptions. One step lower than a scenario is a **cartoon**.)

The Fe–Ni Phase Diagram and the Structure of Iron Meteorites

The striking octahedral structure of iron meteorites (Figures II-6, II-7, and II-8) was originally discovered in the first decade of nineteenth century by E. C. Howard in England, but it is generally associated with the name of

Count von Widmanstätten, who independently discovered it a few years later. The origin of the structure is readily explained by reference to the Fe – Ni phase diagram illustrated in Figure IV-2. The diagram maps three temperature – composition fields. At high temperatures and high nickel concentrations, the only stable phase is γ-iron (taenite). At temperatures below 900°C and low nickel concentrations, α-iron (kamacite) is the only stable phase. Between these fields is a region in which both α and γ phases can stably coexist; at each temperature, the relative amounts of α and γ are inversely proportional to the distances from the bulk composition to the respective phase boundaries. For example, material containing 10% Ni at 600°C is about twice as far from the $\gamma/(\alpha + \gamma)$ boundary as from the $\alpha/(\alpha + \gamma)$ boundary; thus, at equilibrium, the amount of α is twice as great as the amount of γ.

Octahedral patterns were formed when, as a result of cooling, meteoritic

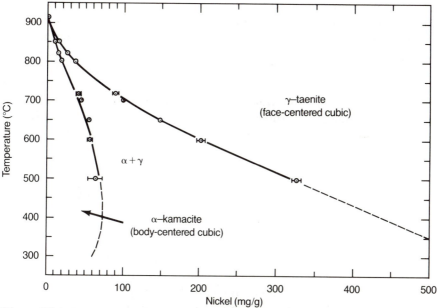

Figure IV-2. Low-temperature portion of the Fe – Ni phase diagram. At high temperatures, γ is the stable phase at all nickel concentrations. Cooling below the $\gamma/(\alpha + \gamma)$ boundary results in the precipitation of α. (Data points with uncertainties shown by bars from J. Goldstein and R. Ogilvie, *Geochim. Cosmochim. Acta* 29:893, 1965; other data points having small uncertainties from M. Hilbert et al., *J. Iron Steel Inst.* 205:539, 1967, and S. Widge and J. Goldstein, *Met. Trans.* 8A:309, 1977; $\alpha(\alpha + \gamma)$ curve below 650°C from J. Willis and J. Wasson, *Earth Planet. Sci. Lett.* 40:162, 1978.

material passed from the γ-stability field into the $(\alpha + \gamma)$ field, and α-iron started to form. Solid phases form three-dimensional structures (crystal lattices), and one can define planes having high atom densities passing through these lattices. The planes are numbered by the inverse of the intercept of the plane with the axes of the unit cell, the smallest repeating unit of the lattice. Thus, a (110) plane intercepts two axes at the edge of the unit cell and is parallel (intercept is at infinity) for the third axis.

A minimum of atomic rearrangement was necessary if one set of crystallographic planes, (110), of the body-centered-cubic α formed parallel to the (111) crystallographic planes in the parental face-centered-cubic phase. There are four possible sets of (111) planes. These are oriented with respect to each other as are the faces of a regular octahedron (only four because opposite sides of a regular octahedron are parallel). These sheets of iron are called lamellae, and the intersections of the lamellae with sections through the meteorite are called bands.

If all iron meteorites remained as equilibrium assemblages down to some fixed temperature (say 750 K), then the thickness reached by lamellae would depend primarily on the amount of α formed—that is, on the bulk nickel concentration as just discussed. In fact, the lowest temperature to which equilibration is approximately maintained depends on the rate at which the metal is cooling; the lower the cooling rate, the lower the "equilibrium" temperature. The phase diagram shows that (for a particular nickel concentration) the lower the temperature, the greater the amount of α relative to γ and, thus, the greater the thickness to which kamacite (α) lamellae can grow. Thus, as mentioned already in Chapter II, the kamacite bandwidth (thickness of kamacite lamellae) increases with decreasing nickel content and decreasing cooling rate. The nickel concentration of kamacite decreases slightly below $\sim 600\,°C$, whereas the nickel concentration of taenite increases monotonically with decreasing temperature.

Iron-Meteorite Cooling Rates

Following the expenditure of the heat source, the rate at which an interior portion of a parent body cooled depended on the burial depth. For iron meteorites formed in cores, the cooling rate depended primarily on the size of the parent body. The approximate relationships between the cooling rate at $500\,°C$, the burial depth, and the parent-body radius are shown in Figure IV-3. Because heat conduction through silicates is much slower than that through metal, the cooling rate is determined by the heat flux through the silicate mantle.

The qualitative considerations regarding kamacite nucleation and growth (see preceding section) have been applied to the development of a

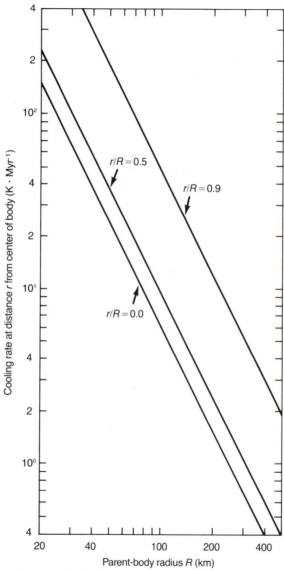

Figure IV-3. For the same initial temperature profiles, cooling rates vary as the inverse square of the parent-body radius R. For samples located $0.0R$ to $0.9R$ from the center, the cooling rates vary roughly as the inverse of the burial depth $(R - r)$. These curves are calculated for planets initially at $\sim 1500°C$ and having a uniform thermal diffusivity of 0.010 cm² · s⁻¹. No heat from radioactivity or other sources was added during the cooling period. The r/R value of 0.5 corresponds to the cooling rate of a core having a mass about 0.23 times the mass of the entire body.

series of increasingly more-sophisticated cooling-rate models. These models are based on the differing rates of nickel diffusion, more rapid in α-iron than in γ-iron. **Diffusion coefficients**[1] in both phases also vary exponentially with temperature, dropping by about a factor of 10 with each 100°C decrease in temperature.

With the electron microprobe, one determines the composition of phases by measuring X rays emitted from a μm-size spot excited by a beam of electrons. Figure IV-4 shows a microprobe nickel trace starting in kamacite and crossing several taenite and kamacite regions before returning to kamacite. The path of the trace is shown by a line on the photo; the kamacite and taenite lamellae are oriented perpendicular to the plane of the section. The dark central part of the broad taenite region consists of plessite, a finely divided mixture of kamacite and taenite. That the nickel concentration gradients are greater in γ than α reflects the lower rate of diffusion in the former and the higher dNi/dT of the $\gamma/(\alpha + \gamma)$ boundary relative to the $\alpha/(\alpha + \gamma)$ boundary.

A numerical simulation of the growth of kamacite and the evolution of nickel profiles in taenite and kamacite is shown in Figure IV-5. Note the growth in kamacite-lamella thickness with time, and the increase in the taenite nickel gradient with time. Note also that, at low temperatures, a low-Ni region develops in the kamacite at the interface with taenite; this results from the decrease in nickel concentration of the $\alpha/(\alpha + \gamma)$ boundary below about 400°C (see Figure IV-2).

Most of the recent cooling-rate determinations have been based on matching the nickel content at the centers of narrow (less than 10 μm) taenite lamellae. The advantage of using narrow lamellae is that they are less affected by minor-element effects or by random variations in nucleation temperatures (the system must normally cool more than ~ 10 K below the $\alpha/(\alpha + \gamma)$ boundary before kamacite nucleation occurs).

An interesting case that is in dispute is the range of cooling rates in IVA iron meteorites. Early papers reported IVA cooling rates negatively corre-

---

[1] The **diffusion coefficient** D is defined by the following equation describing the transport of material down a concentration gradient:

$$\text{flux} = -D \frac{dX}{dx}$$

where X is concentration and x is distance. Typical units are cm for distance, $g \cdot cm^{-3}$ for concentration, $g \cdot cm^{-2} \cdot s^{-1}$ for flux, and $cm^2 \cdot s^{-1}$ for the diffusion coefficient. A similar equation describes the transport of heat down a thermal gradient.

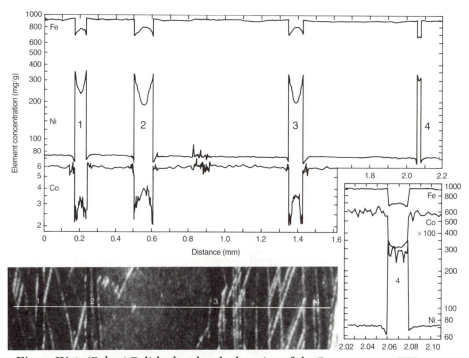

Figure IV-4. *(Below)* Polished and etched section of the Buenaventura IIIB
iron photographed in reflected light. Taenite is black, kamacite is dark gray, and
phase boundaries are white. The parallel white lines inside kamacite are
shock-produced twin lamellae. *(Above)* Microprobe trace of iron, cobalt, and
nickel concentrations along the line marked on the photomicrograph.
Nickel concentration is highest in taenite and lowest in kamacite at the interface
between these phases. Nickel concentrations in some areas (dark on the photo)
fluctuate, reflecting the presence of microscopic grains of kamacite and taenite.
The most-precise iron-meteorite cooling rates are obtained by modeling the
nickel concentration in narrow taenite lamellae such as 4, which is shown in
more detail in the inset. (Photo and diagram by A. E. Rubin.)

lated with nickel content, ranging from 100 K · Myr^{-1} at the low-Ni extreme to ~ 5 K · Myr^{-1} at the high-Ni extreme. The thermal conductivity of
solid Fe – Ni is ~ 30 times greater than that of solid silicates. As a result, a
central core surrounded by silicates would have a nearly constant temperature equal to that of the silicate mantle at the core – mantle interface.

Thus, one expects that all metal in a central core will record the same
cooling rate, independent of distance from the center of the core. If the
mass of the core is 20 to 25 percent that of the parent body, the cooling rate
expected is that given by the $r/R = 0.5$ curve in Figure IV-3. Thus, a large

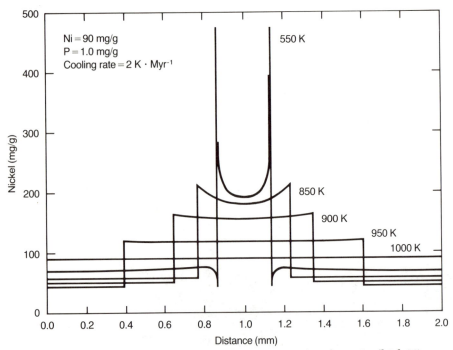

Figure IV-5. Nickel profiles in kamacite (low-Ni, α-iron) and taenite (high-Ni, γ-iron) phases have been simulated at several temperatures for Fe–Ni having a nickel content of 90 mg/g, a phosphorus content of 1 mg/g, and a cooling rate of 2 K · Myr^{-1}. At 1000 K, the system consists entirely of taenite. At 950 K, kamacite has nucleated at the edges of this 2-mm region, and kamacite lamellae are already 0.4 mm thick (note that half of each lamellae is off the diagram.) At 550 K, the kamacite lamellae are ~1.7 mm thick, and strong nickel gradients are present in both phases near the kamacite–taenite interface. (Numerical simulation by K. Rasmussen.)

range of cooling rates in IVA is evidence against a core origin; if these irons formed in a single parent body, they must have originated at a wide range of depths from near surface (high cooling rates) to near center (low cooling rates). This conclusion suggested a **raisin-bread model** — a large number of metal pods distributed throughout the parent body like raisins in raisin bread.

The curious thing about these results is the correlation of cooling rate with composition. If the irons were not part of the same system, one would expect cooling rate to vary randomly with nickel concentration. Further, if irons having different nickel concentrations originated in different "raisins," then (1) why didn't these raisins differentiate? (why isn't a range of compositions found in each raisin?) and (2) what process produced element – Ni correlations in this group that are so similar to those in fractionally crystallized irons? These considerations suggest that the reported differences in cooling rate might not be real but might be an artifact introduced by approximations or errors in the cooling-rate model.

The presence of phosphorus expands the α-stability field toward higher nickel concentrations and increases nickel diffusion coefficients in kamacite. In one recent study, it was reported that allowing for these effects reduces the range of cooling rates and destroys the correlation between cooling rate and composition. Some other recent studies have disputed this finding, and the issue remains open.

About two decades ago, a wide (factor of ~ 10), nickel-correlated range of cooling rates was reported in group IIIAB. These data are now being reinvestigated. As in group IVA, the phosphorus content increases with increasing nickel content, and it seems certain that the cooling-rate range will be reduced by use of phosphorus-dependent phase diagrams and nickel diffusion coefficients. Furthermore, as noted earlier, the large range of iridium contents in IIIAB seems to eliminate all fractionation models except fractional crystallization, and this model requires that all irons form in a single core.

Most published iron-meteorite cooling rates are $\sim 1 \text{ K} \cdot \text{Myr}^{-1}$, corresponding to cores in bodies having radii of about 300 km. This is surprisingly large when one notes that only two asteroids, Ceres and Pallas, have such large radii. Is it reasonable that many (more than 10) large asteroids were destroyed in order to account for the observed compositional spectrum of slowly cooled iron meteorites? Or could the cooling rates be systematically high by a large factor? Could the interpretation that the temperature decrease resulted from the monotonic cooling of a planetary body be incorrect?

Studies of mesosiderites yield results more puzzling still. Typical mesosiderite cooling rates are about $0.1 \text{ K} \cdot \text{Myr}^{-1}$, corresponding to a core in a parent body having a radius of 1000 km, twice as large as that of Ceres, the largest asteroid. Further, as noted in Chapters II and V, the silicate portion of mesosiderites is rich in minerals such as plagioclase that melt at relatively low temperatures. Such basaltic materials are always found on the surfaces of planets such as the Earth and Moon, and there is little doubt that they were once on the surface of the precursor parent body of the mesosiderites. This observation strongly implies that it is incorrect to interpret the mesosiderite data in terms of the essentially constant cooling rates found in planetary cores. It seems more probable that the nickel distributions resulted from the recrystallization of highly shocked small grains during an extensive period at moderate temperatures. During this period, temperatures could have been falling, nearly constant, or fluctuating — without knowing the source of energy, we cannot say which. In any case, the calculated cooling rate would have no direct relationship to burial depth.

Should we also be skeptical about the significance of the iron-meteorite cooling rates? I think that the answer is partly yes, partly no. It is reasonable to expect that the magmatic groups did form in cores, and thus a monotonic cooling seems highly plausible. The geometry of the extensive octahedral areas in iron meteorites is more accurately modeled than that of small, irregular metal grains in mesosiderites (or those in chondrites, for which cooling rates of 0.2 to 10 K · Myr^{-1} have been reported). On the other hand, there may be effects of stress resulting from shock or thermal contraction, and the extrapolation of diffusion coefficients to low temperatures may be in error. I consider it probable that reported cooling rates are systemically low, perhaps by as much as a factor of 10, and that the radii of the parent bodies of the most-slowly-cooled iron meteorites were probably nearer 100 km than 300 km.

Pallasites: Samples of a Core–Mantle Interface

A section through the Salta specimen of Imilac, a typical pallasite, is shown in Figure IV-6. Angular fragments of olivine are separated by a metallic matrix. It appears that a violent event has crushed a rock consisting almost exclusively of olivine and mixed it with molten metal. In some pallasites, the olivine is rounded rather than angular as in the Imilac specimen. The rounding probably resulted from recrystallization of angular material. Thermodynamics favors the elimination of sharp edges and small grains in order to minimize the interfacial area between the olivine and the metal.

Olivine is a refractory silicate in the sense that it is generally the first silicate mineral to crystallize out of cooling silicate liquids having a wide range of compositions (including chondritic). Olivine is the most-common mineral in the Earth's mantle, and it is probably (the possible exception is low-Ca pyroxene) the most-common mantle mineral in the other terrestrial planets and in the parent bodies of the differentiated meteorites. It is thus reasonably probable that (neglecting possible high-pressure phases) the lowest mantle layer of any of these bodies would consist of olivine. Thus the mineralogical composition of the pallasites is precisely that expected for material from a core–mantle interface.

If the core of the pallasite body underwest fractional crystallization, the last liquid to solidify would be expected to be near the core–mantle interface (because of the requirement that the liquid be well mixed—see the following section). Thus, in that case, the metal in the pallasites is expected to have high concentrations of nickel and phosphorus, and low concentrations of iridium. Precisely this is observed in the metal of the main-group

Figure IV-6. Reflected-light photo of a polished and etched surface of the Salta specimen of the Imilac pallasite. The dark angular clasts are olivine, and the lighter matrix is metal. The metal consists of kamacite (white and light-gray band-shaped areas) and plessite (dark-gray, irregularly shaped areas). The olivine is clearly fragmental; thin metal veins commonly separate fragments with complementary surfaces. Pallasites were probably formed by violent events that mixed mantle and core materials. (Smithsonian Institution photo.)

pallasites. The compositions of the main-group pallasites are very similar to those found in the high-Ni, low-Ir extreme of iron-meteorite group IIIAB, and there is a real possibility that these groups originated in the same parent body.

In contrast to the main-group pallasites, three ungrouped pallasites (the Eagle-Station trio) have high iridium concentrations inconsistent with the metal having originated in a fractionally crystallized core. These meteorites also have very unusual oxygen-isotope compositions similar to those in CV or CO carbonaceous chondrites, and far different from those in main-group pallasites. Perhaps these pallasites formed in small melt pools produced by impacts on the surface of a CO-like or CV-like parent body.

What was the violent event that mixed the metal and olivine of the main-group pallasites? The two most-likely explanations are (1) a major impact between the parent body and another large body, or (2) the collapse of the olivine roof into the metallic core due to contraction resulting from cooling. The plausibility of the latter explanation is enhanced by the observation that phase diagrams show that the olivine mantle was solid before crystallization of the core commenced. Because core solidification resulted in a volume reduction of 2 percent, a void could have formed at the core–mantle interface and ultimately have been filled by the collapse of the olivine roof. Conversion of the kinetic energy to heat may have produced some melting of the olivine and metal.

Formation, Evolution, and Fragmentation of the Parental Cores of Iron-Meteorite Groups

As discussed at the beginning of this chapter, it seems probable that the bulk compositions of most iron-meteorite parent bodies were chondritic. If so, most of them had Fe–Ni contents of 80 to 200 mg/g, and FeS (troilite) contents in the chondritic range of about 50 to 150 mg/g.

We will assume an internal origin for the heat source that led to the gravitational separation of immiscible "silicate" and metal–troilite. As the temperature increased, the first "metallic" liquid appeared at 1250 K and consisted of ~ 850 mg/g FeS and ~ 150 mg/g Fe–Ni; the amount of this liquid depended chiefly on the amount of FeS in the starting mixture. If there was enough FeS to permit the melt volume to reach ~ 5 to 8 percent, then this low-melting liquid may have separated to form a molten core. It has been argued that the low sulfur content inferred for the liquids parental to the magnetic iron-meteorite groups are best explained if this early liquid did separate. Continued heating would then produce a denser low-S liquid that would again separate and form a core interior to that of the FeS-rich liquid. During subsequent cooling, the two cores would crystallize independently (the low-S interior core first), with little transport of matter across their mutual interface. In the magmatic groups having low mean nickel contents (less than 80 mg/g), it is likely that interior temperatures approached the melting point of Fe–Ni metal (~1750 K), and that little metal remained in solid form (in either mantle or core) immediately after core formation occurred.

After the heat source subsided, the parent body began to cool. Crystallization of the core could have started in the interior or exterior. In order to achieve the highly efficient fractionation observed in most magmatic groups, it was necessary that the liquid remain well mixed. Diffusion is too slow a process to mix a kilometer-sized core. Stirring can be produced mechanically if, following crystallization at the exterior of the core, blocks of solid Fe–Ni (which is denser than the liquid) break loose and settle to the center. Alternatively, crystallization of the core from the center outward can generate convection both by the liberation of the heat of crystallization and by the rejection of light elements such as phosphorus and sulfur into the liquid if solidification occurred at the bottom of the molten region. The latter mixing mechanism appears more efficient because the former requires either (1) a constant source of stress to break the blocks loose, or (2) "homogeneous" nucleation of the solid directly in the liquid, generally a much less efficient process than "heterogeneous" nucleation on a preexisting crystal.

Because FeS is a major constituent of chondrites, one would expect that the breakup of differentiated asteroids would sometimes produce FeS meteoroids. If the FeS and metal formed a single core, crystallization of about 50 to 90 percent of the core results in FeS becoming a stable solid phase, and an FeS cumulate forms (because of its low density of ~4.7 g · cm^{-3}) at the top of the core, while at the same time Fe–Ni continues to accumulate at the bottom of the molten region. If, as just discussed, a two-layered core forms, FeS may crystallize together with Fe–Ni during the entire crystallization history of the outer, high-S layer.

Meteoroids having high FeS contents are expected to be brittle and thus less resistant to collisional destruction, but one should occasionally fall to Earth. The Soroti meteorite (Figure IV-7) now contains about 60 percent FeS, and the surviving fragments may be more Fe–Ni rich than the meteoroid was prior to atmospheric entry. Biblical reports of rain of fire and brimstone (sulfur) might reflect the occasional fall of FeS-rich meteorites.

The growth of large crystals of γ-iron began during solidification and continued during cooling through the γ field. The largest single crystal ever investigated is visible in the large surface of the Cape York IIIAB iron shown in Figure IV-8. This entire slab having a length of 1.8 m was originally a single γ crystal. The growth of such large crystals appears to require an extended annealing period in the stability field between the solidification

Figure IV-7. The Soroti meteorite consists of FeS (rough texture) and Fe–Ni (areas showing Widmanstätten pattern), as shown by this reflected-light photo. About 60 percent of the meteorite is troilite (FeS), and the amount present in the preatmospheric meteoroid could have been significantly higher. Soroti may be the only example of a meteorite from a FeS-rich cumulate that formed in many asteroidal cores; such meteorites probably are rare because, relative to iron meteorites, they have poor resistance to collisional destruction. (Smithsonian Institution photo.)

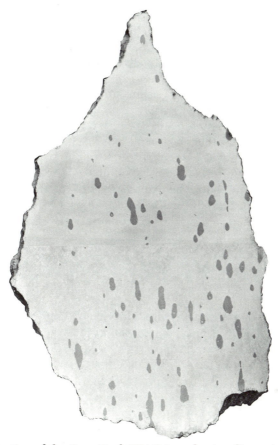

Figure IV-8. Section of the Cape York IIIAB iron, having dimensions of about
1.3 × 1.8 m. The entire slab was a single crystal of γ-iron at high temperatures.
The dark oblong inclusions are troilite (FeS) resulting from the trapping of
liquid during crystallizaton. Low-density phosphates are always in one end (the
"upper" end) of these inclusions, and high-density metal grains are in the other
end, thus defining the direction of the parent body's gravitational field. (Photo
by V. Buchwald.)

temperature in the range from $1200\,^\circ$C to $1400\,^\circ$C and the $\gamma/(\alpha + \gamma)$ boundary (see Figure IV-2).

   After the material cooled into the $(\alpha + \gamma)$ field, the Widmanstätten structure formed by precipitation of lamellae on preferred planes of the precursor γ crystals. As a result of cooling, the core contracted; contraction produced cracks. When at some time a major collision disrupted the parent
body, the core may have fractured along these preexisting cracks. Inclu-

sions of FeS and $(Fe,Ni)_3P$ are also areas where the resistance to fracture was low. The fragments from some of these disruptions were the meter-size meteoroids that eventually reached Earth. In other cases, meteoroids were produced as a result of further impacts breaking up the larger fragments from the primary disruption.

Whenever a body having dimensions of a few meters or less is released into space, its cosmic-ray clock starts to run. As noted in Chapter III, typical cosmic-ray ages of iron meteorites are several hunded megayears. There are many possible fates that await a meteoroid. It can be degraded to dust and gravel as a result of impacts in space. It can be perturbed into an orbit that brings it near Jupiter and, as a result of a close encounter, be ejected from the solar system (that is, be placed in a new "orbit" that is not gravitationally bound to the Sun). It can undergo an impact with an asteroid or with Mercury, Venus, Mars, or Earth. The probability of an Earth impact is a few tens of percent once the meteoroid gets into an Earth-crossing orbit.

The scenario of formation of the nonmagmatic groups IAB and IIICD involves a mechanical enrichment of metal, perhaps by the selective mobilization of metal–FeS melts during impact events. Many details regarding the generation of the chemical fractionations in these groups remain obscure, but a key factor may be the differences in the temperatures of the shock-generated melts. The later histories of these groups parallel those of the members of magmatic groups starting with the period of extensive annealing in the γ field and continuing through capture by the Earth.

Suggested Reading

Buchwald, V. F. 1975. *Iron Meteorites.* University of California Press. A wealth of descriptive material regarding iron meteorites, including excellent photos and detailed historical and metallographic notes on 480 irons.

Goldstein, J. I., and H. J. Axon. 1973. The Widmanstätten figure in iron meteorites. *Naturwissenschaften* 60:313. A theoretical discussion of the formation of the Widmanstätten pattern and of the evidence supporting raisin-bread models for certain groups.

Kelly, W. R., and J. W. Larimer. 1977. Chemical fractionations in meteorites, VIII. Iron meteorites and cosmochemical history of the metal phase. *Geochim. Cosmochim. Acta* 41:93. Detailed discussion of the formation of iron meteorites, with emphasis on nebular condensation and partial melting in parent bodies.

Kracher, A., and J. T. Wasson. 1982. The role of S in the evolution of the parental cores of the iron meteorites. *Geochim. Cosmochim. Acta.* 46:2419. A technical discussion of the role of sulfur in the evolution of the iron-meteorite parent bodies.

Scott, E. R. D. 1972. Chemical fractionation in iron meteorites and its interpretation. *Geochim. Cosmochim. Acta* 36:1205. A review of elemental concentrations in iron meteorites, with emphasis on the evidence for fractional crystallization.

Scott, E. R. D. 1979. Origin of iron meteorites. In *Asteroids,* ed. T. Gehrels (University of Arizona Press), p. 892. A brief technical review of the properties of iron meteorites and ideas about their origins.

Wasson, J. T. 1972. Parent-body models for the formation of iron meteorites. *Proc. 24th Intern. Geol. Cong.,* Sect. 15, p. 161. A brief discussion of the formation of iron meteorites in parent bodies.

Wood, J. A. 1968. *Meteorites and the Origin of Planets.* McGraw-Hill, Chapter 3 discusses the determination of iron-meteorite cooling rates.

CHAPTER V

Igneously Formed Silicate-Rich Meteorites

Because we live on a differentiated planet, we have rather clear ideas about the basic processes associated with melting of rocks and with the inverse process, crystallization. Numerous laboratory experiments have been conducted to simulate these processes. For mixtures of the most common silicate minerals, we can predict the composition of the melt and predict which phases will remain as solids as a function of temperature. In this regard, it is important to note that only pure substances and a few mixtures called eutectics have sharp melting temperatures.

Although a few spectacular rock bodies were formed by fractional crystallization, most terrestrial surface **igneous** rocks (rocks formed by melting) are produced by the process called **partial melting,** in which a melt forms in equilibrium with several solid phases, then is removed and (typically) quickly chilled. The general name for rocks formed in this way is **basalt,** and basalt is the most common type of rock on the Earth's surface.

Other rocks (especially some having high contents of olivine) formed as unmelted residues and thus are the complements of basalts and other rocks that formed as partial melts. (See Appendix C for formulas of olivine and other minerals commonly found in meteorites.)

On Earth, igneous rocks can have much-more-complex origins involving the mixing of unrelated magmas, partial melting followed by fractional crystallization of the separated liquid, remelting of sediments, and so on. Some of these processes also occurred on the meteorite parent bodies but, because of the absence of processes involving water, the sets of processes forming the differentiated silicate-rich meteorites were considerably simpler. (If originally present, water vapor would have readily escaped from the parent body's gravitational field during heating to igneous temperatures and a short period of igneous activity — perhaps 200 to 300 Myr versus 4.5 Gyr for the Earth.)

Origin of the Igneous Clan

Rocks that crystallized from a melt and were subsequently stored at relatively low temperatures generally have textures that testify to their igneous origins. The Allan Hills A78158 clast shown in Figure V-1 is an example. The laths of plagioclase clearly crystallized while liquid was still present; their long, needle or lathlike shapes could be produced only if their growth were unimpeded by other solid phases. Laboratory experiments in which

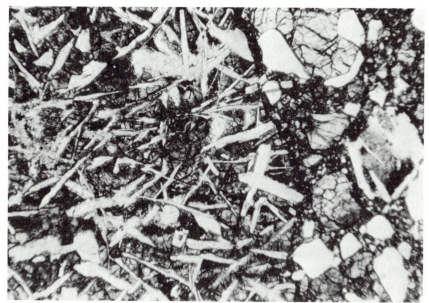

Figure V-1. This section of Allan Hills A78158 eucrite was photographed in transmitted light. In the clast that occupies the left two-thirds of this view, needles of white plagioclase cross pigeonite crystals; the plagioclase must have grown while much of the material was still liquid. This structure demonstrates the origin of A78158 by crystallization from a melt. The microbreccia on the right is compositionally similar to the igneous clast. Width of the photo is 1.6 mm. (Photo by R. Hewins.)

material having a composition similar to this clast is melted in a crucible, then slowly cooled, show that the pyroxene started to crystallize at about the same time but, because it forms equidimensional crystals, most of the pyroxene ends up between the long plagioclase crystals.

Most differentiated stony (silicate-rich) meteorites belong to the igneous clan consisting of the eucrites, howardites, mesosiderites, diogenites, and probably also the pallasites. Similar textures are found in the two largest groups—the **eucrites** (20 falls) and **howardites** (18 falls)—and in the smaller group of **mesosiderites** (only 6 falls, but some of the 14 finds are relatively unweathered and suitable for detailed study). As noted in Chapter II, the eucrites (with rare exceptions) are monomict breccias (clasts and matrix are all from the same rock), whereas howardites and mesosiderites are polymict breccias of regolithic derivation. Howardites and the silicate portions of mesosiderites are mainly basaltic in composition, but only a few rare clasts have retained unaltered basaltic textures. The howardites also contain pyroxene-rich clasts similar to diogenites, and the mesosiderites contain both pyroxene-rich and (rare) olivine-rich components.

There is evidence that the eucrites formed as low-degree partial melts of a relatively primitive source material. One observation indicating that the source had not suffered significant fractionation is provided by the nearly unfractionated (relative to chondrites) patterns of **rare-earth elements**[1] illustrated in Figure V-2. The rare earths are **trace elements** (elements having concentrations too low to allow formation of their own mineral phases, so that instead they substitute for the more-common elements in the mineral sites of those elements). The rare earths are also **incompatible elements,** so called because their sizes are incompatible with their substitution for the common mineral ions of similar ionic charge. During partial melting, the concentration of most of the incompatible elements in the melt is much greater than that in the solids that remain unmelted.

If the eucrites are partial melts and if the material parental to the eucrites had rare-earth abundances similar to those in ordinary chondrites, then the maximum degree of partial melting is the inverse of the maximum rare-earth abundance shown in Figure V-2. Thus, for the Antarctic meteorite Y74450, the maximum degree of melting is $\frac{1}{15}$, or 0.067. Because some (albeit small) portion of the rare earths remained in the solid, the maximum amount of melting involved if Y74450 was formed by a "primary" partial melting of the bulk planet is ~6 percent. These simple arguments cannot be applied to the howardite and mesosiderite data because of the dilution of the "partial-melt" material by other rock types having high contents of ferromagnesian silicates. We shall see that the formation of the mesosiderites may have involved other complications.

In Figure V-2, the samples having high rare-earth concentrations typically have europium values 10 to 50 percent lower than those of the neighboring elements, samarium and gadolinium. Generally, this fact indicates that the melt was in equilibrium with a significant quantity of anorthite.[2] In

---

[1] The 14 (actually 15, but all isotopes of promethium are radioactive) **rare-earth elements (REE)** are very similar in their chemical behavior. In this portion of the periodic table, increasing the atomic number adds electrons to the $4f$ electron shell, and these $4f$ electrons do not participate in chemical reactions. As a result, the chemical behavior of the rare earths is mainly controlled by the "valence" (bond-forming) electrons, two in the $7s$ shell and one in the $5d$ shell. In most meteoritic, terrestrial, and lunar rocks, 13 of the rare earths (europium is the exception) are in the $+3$ ionic state as a result of losing these electrons and forming ionic bonds.

The natural abundances of the rare earths vary by a factor of 30 (highest in cerium, lowest in thulium and lutetium). It simplifies comparisons to remove these element-to-element variations by dividing rare-earth concentrations in rock samples by those in ordinary chondrites, as in Figure V-2.

[2] Europium is a unique rare earth. In all lunar (and in some meteoritic and terrestrial) rock systems, a major fraction of the europium is present in the $+2$ state. Europium's "third" valence electron enters not the $5d$ orbital (as in other rare earths) but rather a $4f$ orbital, which then becomes exactly half-full. This is a partic-

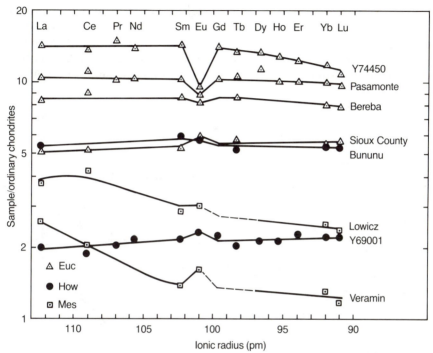

Figure V-2. Ordinary-chondrite-normalized rare-earth concentrations in eucrites, howardites, and mesosiderites. These values mainly reflect the strong partitioning of these elements into the partial melts that crystallized to form the basaltic components of these rocks. Eucrites and howardites show relatively flat, unfractionated patterns suggesting high degrees (~10 percent) of partial melting. Mesosiderite patterns show an enrichment in light rare earths (La, Ce) relative to heavy (Yb, Lu), attributed to low degrees (~3 percent) of partial melting. The rare-earth elements (except europium) are almost entirely in the +3 oxidation state. The large portion of europium that is in the +2 state efficiently substitutes for calcium in anorthite; thus, low europium ratios reflect the presence of appreciable anorthite in the solid residue in equilibrium with the basaltic melt, whereas high europium ratios indicate that the amount of anorthite in the final rock was enhanced during crystallization or impact-induced mixing. (Data for Y74450, Pasamonte, and Y69001 from H. Wänke et al., *Proc. Lunar Sci. Conf. 8th*, p. 2191, 1977; data for Bununu, Lowicz, and Veramin from D. Mittlefehldt et al., *Geochim. Cosmochim. Acta* 43:673, 1979; data for Bereba and Sioux County from D. Mittlefehldt, *Geochim. Cosmochim. Acta* 43:1917, 1979.)

ularly stable electron configuration, and thus more energy is required to remove the third electron from europium. The portion of europium in the +2 state behaves differently from the +3 rare earths. It is no longer incompatible but substitutes efficiently for calcium in anorthite, $CaAl_2Si_2O_8$.

contrast, howardites and mesosiderites often have europium concentrations slightly higher than samarium and gadolinium concentrations, as expected if these breccias contained a minor anorthite-rich component.

Although the rare earths other than europium are generally incompatible, they are not identical in behavior but show systematic differences in

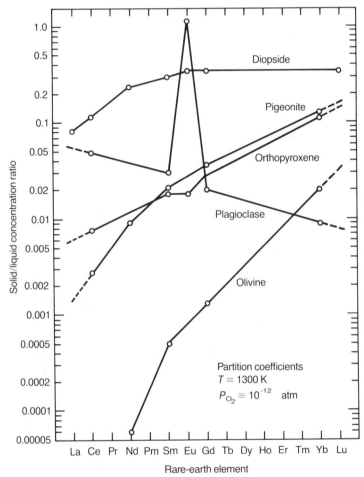

Figure V-3. Solid–liquid partition coefficients for a basaltic magma in equilibrium with five possible source minerals. The +3 rare earths are excluded from these minerals because they cannot substitute effectively for the major elements. Such excluded elements are designated incompatible elements. Most europium is in the +2 oxidation state that can substitute for calcium in diopside. (Data from G. McKay, *Lunar Planet. Sci.* 13:493, 1982; G. McKay and D. F. Weill, *Proc. Lunar Planet. Sci. 8th*, p. 2339, 1977; and other sources cited therein.)

their degree of uptake by specific minerals. Solid–liquid partition ratios for olivine, orthopyroxene, diopside, and plagioclase are illustrated in Figure V-3. The high partition ratio of europium into anorthite and its depression in the other three minerals reflects the fact that much of the europium in the melt is in the $+2$ oxidation state at the experimental O_2 pressure of $\sim 10^{-13}$ atm (a value appropriate to the eucrite parent body); partition ratios for the other REE never exceed 0.35 in diopside, or 0.15 in the other minerals. In olivine, orthopyroxene, and diopside, partition ratios increase with increasing atomic number, whereas the inverse is true for plagioclase. In some minerals, partition ratios for europium are depressed, reflecting a lower ratio for the portion in the $+2$ oxidation state.

The REE patterns in the eucrites having the highest concentrations show a decreasing trend for the heaviest rare earths (Figure V-2). In Y74450, the lutetium concentration is about 0.75 times the mean concentration of the La–Sm group. The "missing" rare earths were apparently taken up by one or more phases having a preference for heavy rare earths. The Gd–Lu depletion increases with increasing atomic number, thus it appears that diopside was not the dominant phase, because diopside partition coefficients are essentially constant throughout the Gd–Lu range. Instead, a mineral such as orthopyroxene seems required; if olivine were the dominant residual solid, then the low lutetium abundance would require very small degrees of partial melting.

Although dilution by other components prevents firm conclusions about the degree of partial melting in the basaltic components of the howardites and the mesosiderites, examination of Figure V-2 shows that the patterns typical of each group are distinctly different. Essentially no fractionation can be resolved in the howardite patterns, whereas light rare earths are about twice as abundant as heavy rare earths (Yb, Lu) in the mesosiderites. The simplest model of mesosiderite origin is that a metallic core fell onto a similar-sized body covered by basaltic materials, and one might guess that these largely basaltic materials were identical to howardites. The rare-earth data are inconsistent with such a model, and there are other compositional differences indicating that this model is incorrect. Relative to average howardites, average mesosiderites have slightly higher amounts of "diogenitic" elements such as Mg, Cr, and V and slightly smaller amounts of "eucritic" elements such as Na, Ca, Ti, and Sm. The most plausible models of the mesosiderite rare-earth patterns indicate low (~ 3 percent) degrees of partial melting of a source that had previously lost a partial melt. This is a more complex planetary model than that required to explain the eucrite and howardite patterns. In fact, as we will discuss later, formation of pallasites and eucrites in the same body requires that eucrite formational processes be more complex than the simplest set capable of explaining the rare-earth data.

Figure V-4 is a petrological **ternary phase diagram.** Each apex corresponds to 100 mol% of the component listed there. In order to plot rock compositions on such a diagram, the amounts of the three components must be normalized to make them total exactly 100 percent. Any other components that may be present are usually ignored. The V-shaped tick marks along the axes correspond to increments of 5 percent; each arm of the V points towards the corresponding mark on one of the other axes. As an

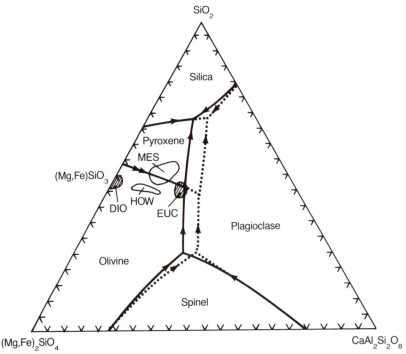

Figure V-4. Olivine–anorthite–silica pseudoternary diagram showing phase relations between melts and solids. The dotted curves apply to liquid having FeO/(FeO + MgO) ratios of 0.30 (from D. Walker et al., *Earth Planet. Sci. Lett.* 20:325, 1973). The solid curves apply to liquid FeO/(FeO + MgO) ratios of 0.60 (in the pyroxene–olivine–plagioclase region, these are taken from E. Stolper. *Geochim. Cosmochim. Acta* 41:587, 1977; in the other areas, these are interpolated from the Walker et al. study). The eucrite field falls at the position expected for a liquid having an FeO/(FeO + MgO) ratio of 0.6 in equilibrium with pyroxene, olivine, and plagioclase, as expected if the eucrites formed by partial melting of a solid containing these three phases. See text for more details. (Meteorite data from A. Simpson and L. Ahrens, *Comets, Asteroids, Meteorites,* University of Toledo Press, 1977. p. 445, and other Ahrens group papers cited therein.)

exercise in reading the diagram, note that the point "35% olivine, 45% SiO_2, 20% $CaAl_2Si_2O_8$" lies just left of the center of the shaded eucrite (EUC) field. Also note that in this diagram orthopyroxene $(Mg,Fe)SiO_3$ is on the left axis, halfway between olivine and SiO_2.

The actual diagram shown here is a **pseudoternary liquidus diagram.** *Ternary* refers to the three end components. The prefix *pseudo* refers to the fact that it is based on a real rock system containing additional components (such as albite, $NaAlSi_3O_8$, and chromite, $FeCr_2O_4$) that could not be included on the diagram. Phase boundaries on the true ternary would differ somewhat from those shown here. The petrological term **liquidus** refers to the composition and temperature of a liquid in equilibrium with one or more solids having compositions that are generally near the axes or apices of the diagram.

Most of the curves on the diagram are marked with single arrows. These curves give the compositions of liquids in equilbrium with the components having stability fields on either side of the curve. For example, the roughly vertical curve just below the center of the diagram gives the composition of the liquid simultaneously in equilibrium with olivine and plagioclase. The arrows point in the direction the liquid follows with decreasing temperature. Cooling of a liquid on such a curve results in the precipitation of each of the equilibrium phases and causes the composition of the liquid to move in the indicated direction. The curve between the pyroxene and olivine fields marked by two arrows is somewhat different; cooling of a liquid lying on this curve results in the simultaneous dissolving of olivine and precipitation of pyroxene.

Two sets of curves are shown in Figure V-4. The dotted curves were first devised by David Walker and colleagues to explain phase relationships found in lunar highlands rocks. They correspond to a FeO/(FeO + MgO) mole ratio of 0.30 in the liquid. The solid curves in the olivine–pyroxene–plagioclase region are based on studies of eucrites by Edward Stolper; they correspond to the FeO/(FeO + MgO) ratios of approximately 0.6 observed in eucrites. The solid curves in the remainder of the diagram are extrapolations of the curves derived for lunar rocks.

The lowest temperatures at which liquids can be in equilibrium with specific components are at the low-temperature ends of the curves involving those components. Thus, when a solid containing olivine, pyroxene, and plagioclase is heated, the first liquid to form would have a composition at the intersection of the pyroxene–olivine and olivine–plagioclase liquidus curves. The coincidence of the eucrites (actually only the so-called noncumulate eucrites) with this composition is consistent with their having been formed by partial melting of a source containing these three components (and, after melting, *still* containing some residual amount of each).

Textural evidence (Figures II-5 and V-1) shows that eucrites cooled rapidly; thus, after production by partial melting, the liquid was brought into an environment where heat could be rapidly dissipated. These basalts were probably extruded as a thin sheet onto the surface of the parent body, where they chilled rapidly by radiating their heat into space. Other more-complex sets of igneous processes could also produce liquids near the eucrite composition but would have produced a sequence of other materials not observed to fall as meteorites. Partial melting is considered more plausible than these models because the only liquid it yields is eucritic in composition. Differences among eucrites — such as the variable rare-earth contents illustrated in Figure V-2 — are attributed to slight differences in the degree of melting and/or minor degrees of fractional crystallization (see Chapter IV) or the settling out of crystals during cooling.

There is good reason to believe that the diogenites, howardites, and eucrites formed on the same parent body. In howardites, clasts that are indistinguishable from eucrites coexist with clasts indistinguishable from diogenites. Howardite compositions, both in terms of major (Figure V-4) and minor elements, can be precisely modeled by mixing eucritic and diogenitic components. Some meteorites that are generally designated diogenites consist dominantly of orthopyroxene and have minor eucritic components. These "diogenites" are thus transitional to howardites (Johnstown is an example).

As indicated in Figure V-5, the three groups are indistinguishable on a $\delta^{17}O - \delta^{18}O$ diagram. These three groups fall together with mesosiderites and main-group pallasites in a small field near the terrestrial fractionation line but about 2‰ lower than the mean $\delta^{18}O$ value for the Earth. In fact, these members of the **igneous clan** have the lowest $\delta^{18}O$ values found among those chondrite and differentiated-meteorite groups that cluster about the terrestrial–lunar field, and that appear to have formed in the inner solar system (see Chapter IX).

The intimate mixing of eucritic and diogenitic material in the howardites implies that the two end-component groups originated in adjacent regions on the parent body. If we accept the arguments indicating that the eucrites formed as partial melts, then the simplest model for the diogenites would be that they are the residual material from which the eucritic material was extracted. In fact, there are several good reasons for rejecting this hypothesis. First, as indicated by the position of their field in Figure V-4, the diogenites contain little-to-no plagioclase. This fact conflicts with the observed europium depletions in many eucrites, which appear to require larger amounts of plagioclase in the residue. Another problem is the small-to-negligible amount of olivine in the diogenites, an amount that is smaller than expected if, as discussed earlier, eucrites formed at the intersection of the olivine–pyroxene and plagioclase–olivine fields in Figure V-4. A third problem may be the most serious. (FeO/(FeO + MgO) ratios in diogenites

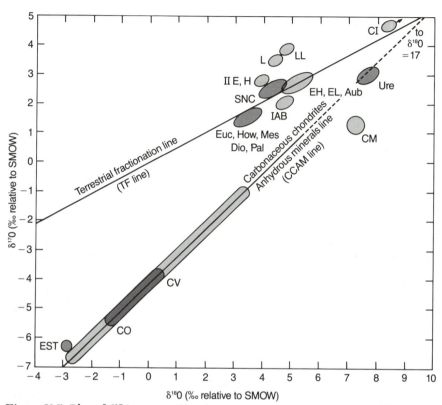

Figure V-5. Plot of $\delta^{17}O$ versus $\delta^{18}O$ showing the locations of differentiated silicate-rich meteorite groups (stippled) and, for reference, the locations of the ten chondrite groups (lined). The δ notation gives $^{17}O/^{16}O$ and $^{18}O/^{16}O$ ratios; see the discussion of Figure III-9 for more details. Five differentiated groups (Euc, How, Mes, Dio, Pal) share a common field just below the terrestrial fractionation (TF) line; this igneous clan formed under closely related conditions, probably in a single parent body. The Shergotty, Nakhla, and Chassigny (SNC) differentiated meteorites form a field just above the TF line. The field shown for the EH and EL chondrites and the aubrites also includes the terrestrial field; like the igneous and SNC clans, these meteorites probably formed in the inner solar system (see Chapter IX). In contrast, the ureilites fall in a sparsely populated region away from the "inner-solar-system" cluster, perhaps reflecting formation farther from the Sun. (Data mainly from R. Clayton et al., *Earth Planet. Sci. Lett:* 30:10, 1976.)

are 25 to 27 mol%, whereas those in eucrites are 50 to 67 mol%, as listed in Table II-2. Although the fact that the ratio is higher in eucrites is consistent with the observation that melts always have higher FeO/(FeO + MgO) ratios than do residues containing pyroxene or olivine, the difference is too large to explain in terms of a single-step process. Laboratory experiments show that a melt in equilibrium with orthopyroxene having an FeO/(FeO + MgO) of 26 mol% should have a ratio of about 45 mol%, well outside the eucrite range.

The conclusion that eucrites and diogenites probably originated on the same body but are not complementary differentiates is important because it eliminates the following appealingly simple model that was popular two decades go: (1) a chondritic body was heated to just the right temperature needed to generate a eucritic partial melt; (2) the melt included the bulk-heat-producing elements potassium, uranium, thorium and (if heat-producing) aluminum; (3) extrusion of the melt led to rapid cooling and, as a result of the removal of the heat-producing elements from the interior, the end of igneous activity; (4) the unmelted residue consisted of diogenites; (5) the howardites formed by the impact-generated mixing of eucrites and diogenites. More-complex igneous models clearly are required, apparently involving more than one episode of melt generation and possibly including fractional crystallization as well as partial melting.

As discussed earlier, the relatively unfractionated patterns and the enrichments of rare earths by factors of 5 to 16 in eucrites are consistent with formation by the partial melting of unfractionated chondritic material. If this model were correct, then the FeO/(FeO + MgO) ratio in the chondritic parental material was at least as high as 35 mol%. This possibility is effectively ruled out by the low Fe/Mn atom ratio of ~35 in the eucrites and similar values in the howardites (~33) and diogenites (~31). During igneous processes, manganese closely follows iron; the Fe/Mn ratio in coexisting melt and solid are nearly the same. The only significant exception to this rule involves olivine, for which available data suggest that $(Fe/Mn)_{solid}/(Fe/Mn)_{liquid} \cong 1.1$ to 1.4. As a result of this coherence between iron and manganese, partial melting of a chondritic parent should lead to eucrites having Fe/Mn atom ratios essentially identical to the FeO/MnO mole ratio in the parental chondritic material (the iron present as metal and FeS does not enter the silicate melt).

Excluding the EH and EL chondrites (much too reduced) and the CI and CM chondrites (much too oxidized), the total range of the FeO/MnO mole ratios in the chondrite groups that might resemble the eucrite parents is from 32 in the H group to 96 in the CV group. If we narrow the list to the three groups LL, CO, and CV having FeO/(FeO + MgO) ratios near 35 mol%, then the FeO/MnO range is from 57 to 96, or from 1.5 to 2.5 times as great as eucritic values. The eucritic FeO/MnO ratio is much more consistent with formation from a chondritic material having a degree of oxidation similar to that of the H-group chondrites.

If the main-group pallasites formed in the same parent body as the eucrites, then their formation also requires an FeO/(FeO + MgO) ratio in the starting material roughly in the H-group range. The pallasites consist of mixtures of olivine and metal with a negligible content of other silicates. As discussed in Chapter IV, the juxtaposition of metal and refractory silicate suggests that these are samples of a core – mantle interface, and other evidence is consistent with such a model. Formation of a pure olivine layer at the bottom of the mantle requires temperatures high enough to completely melt the other silicate minerals. If the starting material were chondritic, most (more than 90 percent) of the lithophile elements would be in the melt at a temperature of ~ 1600 K, and the FeO/(FeO + MgO) ratio of the melt should be the same as or slightly higher than that in the starting material. From laboratory studies, we can infer that a mean FeO/(FeO + MgO) ratio of 11 to 14 mol% in the pallasitic olivine requires about 17 to 22 mol% in the melt, a value consistent with a starting FeO/(FeO + MgO) ratio about the same as that observed in H-group chondrites.

Because the pallasite FeO/(FeO + MgO) ratio and the eucrite FeO/MnO ratio both imply that the starting composition of the eucrite parent body resembled the H-group chondrites in degree of oxidation, we should inquire how such a starting material could produce the very high FeO/(FeO + MgO) ratios observed in the eucrites. The simplest scenario is the following: following the formation of the pallasites, the molten portion of the parent body continued to cool, resulting in the precipitation of olivine (perhaps later together with pyroxene). The precise course of the crystallization cannot be predicted with precision but, if olivine were the only phase that crystallized, calculations show that the FeO/(FeO + MgO) ratio of the liquid can be increased to 35 mol% by crystallizing ~ 60 percent of the mass as olivine, and that this amount of crystallization would leave rare-earth patterns that are still essentially flat. Thus, chilling this residual liquid would produce a suitable solid material that could yield a eucritic liquid if partially melted at a later time. Alternatively, the fractional-crystallization process may have continued, and the eucrites formed from the partial melting of material from several outer layers of the body, perhaps previously mixed by impacts. In this case, the diogenites might represent material produced during the primary crystallization, and howardite and mesosiderite breccias could include as components both primary and secondary differentiates. Obviously, these ideas are highly speculative, but they are nonetheless illustrative of reasonably plausible sets of circumstances that could have produced all members of the igneous clan in a single parent body.

Meteorites belonging to the igneous clan have low concentations of volatiles relative to those observed in terrestrial rocks and in chondrites; the volatile contents of lunar rocks are similar to those in the igneous clan. As an example, the typical potassium content of a eucrite is about 400 $\mu g/g$, whereas typical values in terrestrial basalts are about 8000 $\mu g/g$. To make the comparison with chondrites, we must allow for the fact that potassium is incompatible to about the same degree as the rare earth lanthanum. In chondrites (excluding a few chondrites that have experienced melting), the K/La weight ratio varies from 800 to 2400; that in terrestrial basalts is about 450, in eucrites about 130, and in lunar rocks about 70.

The two chief hypotheses to account for the low volatile contents in the eucrites and related meteorites are these: (1) the volatile loss resulted from processes occurring in the solar nebula prior to the formation of the parent body; or (2) the loss occurred as a result of igneous processes occurring in (and on) the parent body. The first hypothesis amounts to a proposal that some chondritic materials formed with a K/La ratio about 10 times lower than is known in the chondrites in our meteorite collections. Because the known chondrites provide the only sure evidence of the fractionations that occurred in the solar nebula, I doubt that this hypothesis is correct. A more-detailed version of the second hypothesis involves the repeated eruption of magmatic volatiles during the period when the parent body was molten; if the mean velocity of the dominant magmatic gaseous species was above the escape velocity, then the entire gas cloud would escape into space upon rupture of the chamber. The parent-body radius was probably about 100 km, and surely was not greater than 500 km. According to Equation IV-1, escape velocities were thus about 140 to perhaps 700 m $\cdot$ s^{-1}. It is probable that the dominant gaseous species was H_2O, but CO and CO_2 are also possibilities. Gas-kinetic theory shows that the mean velocity v of the gas in m $\cdot$ s^{-1} is

$$v = \left(\frac{8RT}{\pi M}\right)^{1/2} \qquad (V\text{-}1)$$

where R is the gas constant (8.31 J $\cdot$ K^{-1} $\cdot$ mol^{-1}), T is the absolute temperature, and M is the molecular weight in kg $\cdot$ mol^{-1} of the gaseous species. For $v = 700$ m $\cdot$ s^{-1}, this formula yields $T = 416$ K, 648 K, and 1018 K for H_2O, CO, and CO_2, respectively. Although some cooling as a result of expansion during eruption will occur, it seems probable that the highest of these temperatures is still on the low side of mean temperatures of eruptive gases during a period when much of the parent planet was molten, and thus that magmatic volatiles could escape from the largest known asteroid if they were able to reach the surface. Because the high impact rate during the first 100 Myr of solar-system history would have produced repeated fracturing of the solid crust and thus eruption of magmatic gases, the second hypothesis seems quite plausible.

Differentiated Silicate-Rich Meteorites from Mars?

Two classes of meteorites once linked to groups in the igneous clan are now known to be distinct, and they almost certainly formed in independent parent bodies. Early publications suggested that the Shergotty basalt is an unusual species of eucrite, but this hypothesis is ruled out by its distinctly different oxygen-isotope composition (Figure V-5) and a content of volatiles significantly higher than that of eucrites. The higher volatile content is particularly interesting in the light of the arguments just presented for believing that eucritic volatiles were lost as a natural consequence of extensive melting of the igneous-clan parent body. If those speculations are correct, Shergotty formed on a body too large to allow volatiles to escape or under conditions leading to heating and cooling processes too rapid to lead to outgassing. Shergotty is highly shocked, and one speculation associates this shock with removal from a very large parent body — Mars!

The four Shergotty-class meteorites are chemically and isotopically linked to a meteorite called Nakhla and two similar stones and to Chassigny, which forms a third class of meteorite. The Nakhla-class meteorites consist dominantly of augite (essentially diopside with a moderate aluminum content); Chassigny consists mainly of olivine. As shown in Figure V-5, oxygen-isotope compositions are very similar in the members of the three classes. Another link is a common radiometric formation age of 1.3 Gyr, much lower than the values of 4.4 to 4.6 Gyr measured in meteorites of the igneous clan. The three classes also share high volatile contents, including hydrated minerals in the Nakhla class and in Chassigny. These three related classes are often called the SNC (pronounced "snick") meteorites.

The high volatile contents suggest origin on a large parent body with a gravitational field great enough to retain volatiles. The low igneous ages also suggest origin on a body large enough to efficiently retain its internal heat and maintain temperatures near melting for at least 3.2 Gyr; a body larger than the Moon seems required. Studies of rare gases in Shergotty-class stones show high $^{129}Xe/^{132}Xe$ ratios consistent with retention of radioactive ^{129}Xe in an atmosphere, and with the trapping of some of this atmosphere in the meteoritic material during the shock event. Also observed are enhanced contents of those rare-gas isotopes that can be produced by neutron capture in rocks, consistent with the fact that the thin Martian atmosphere permits a relatively high cosmic-ray flux at the surface.

The Shergotty-class meteorites were shock metamorphosed about 180 Myr ago; this is the suggested time of their release from the parent body. Because their cosmic-ray exposure age is only ~2.5 Myr, however, the block released from the parent body must have had linear dimensions of ~ 10 m, such that much of the interior was shielded from cosmic rays (as noted in Chapter III, the mean penetration depth of cosmic rays is ~ 1 m). Although the similar formation ages suggest derivation of all SNC meteorites from the same igneous complex, their petrological diversity suggests origins at widely separated (much more than 100 m) regions within the complex. Thus it seems doubtful that all could have been contained in a single 10-m block.

The key unresolved question is whether an impact could eject ≥ 10-m blocks from Mars with velocities in excess of the escape velocity of 5 km $\cdot$ s^{-1}. Laboratory studies of impacts indicate that high velocities are imparted to only a tiny fraction of the ejecta; this high-velocity ejecta is comminuted to small sizes and is generally heavily shocked. There is some evidence that Mars has much near-surface H_2O present as permafrost. One suggestion is that the impact energy evaporated much of the H_2O in the crater, and that this sudden evaporation generated a huge expanding cloud that entrained some large blocks, raising them to the escape velocity. Presumably each class was in a separate block. My assessment is that it is ~ 50 percent likely that the SNC meteorites originated on Mars, but that confirmation will be possible only when samples from Mars have been returned for analysis in terrestrial laboratories.

The Eagle-Station Trio

The stippled field of a cluster of three ungrouped pallasites is labeled EST in Figure V-5. The largest of these pallasites is Eagle Station, and a common designation for the three is the Eagle-Station trio. The separation from the field of the main-group pallasites (see Chapter IV) confirmed compositional evidence (Ga/Ge ratios 20 times lower than main-group ratios) for designation as a separate class. The oxygen-isotope data suggest that the Eagle-Station trio formed by melting and igneous fractionation of material similar to the CO chondrites. If, as suggested in Chapter IX, the CO material originated in the outer solar system (more than ~ 3 AU from the Sun), then even in this region some parent bodies were heated to temperatures that could produce differentiation. If the heat source in the inner solar system was in appreciable part solar (a superluminous Sun or induced electrical currents), then the heating of a CO body may have been largely by impacts, and these pallasites may not be samples of a core–mantle interface.

The Ureilites

Figure V-6, a photograph of a thin section of the Goalpara ureilite, illustrates the features common to most ureilites. The most abundant phase is

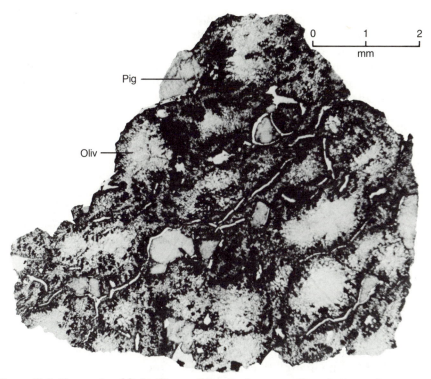

Figure V-6. Transmitted-light photo of the Goalpara ureilite, showing the characteristic ureilitic features: light-gray polycrystalline olivine domains, slightly darker pigeonite grains with sharply defined boundaries; opaque carbon- and metal-rich regions, and white cracks (most of them oriented SW–NE in this section). The cracks commonly border pigeonite grains. (Smithsonian Institution photo; British Museum thin section no. 51187.)

olivine, generally present in polycrystalline domains having dimensions of about 0.5 to 1 mm. The only other common silicate is low-calcium clinopyroxene in Haverö, pigeonite in the other ureilites. Long, oriented cracks and voids are present in ureilites other than Kenna. Dark veins containing graphite, diamonds, and Fe–Ni metal border the cracks, but they are found also in areas not showing obvious cracks.

The cracks, the diamonds, and the polycrystallinity of the olivine appear to have resulted from shock. Shock pressures are usually expressed in GPa, where 100 kPa $\cong$ 1 atm. The ureilites experienced shock pressures in the range from 30 to 80 GPa, among the highest values recorded in meteorites.

The oxygen-isotope composition of ureilites (Figure V-5) places them outside the highly populated area surrounding the Earth–Moon field, in a region near the CM-matrix composition but having a $\delta^{17}O$ value about 1‰

higher. The only meteorite having a similar oxygen-isotope composition is a carbonaceous clast in the Bencubbin meteorite.

The ferromagnesian silicates apparently were produced either as residues from extensive fusion or as early cumulates from fractional crystallization near the base of a magma chamber of unknown size and depth. The pigeonite and other accessory mineral phases could have formed from some of the associated liquid that became trapped between olivine grains. Although additional complementary rock types must have been produced, the available oxygen-isotope data indicate that none have fallen as meteorites.

The carbon-rich vein material must have been introduced later, because it could not have existed in equilibrium with the relatively oxidized ureilitic materials during their igneous formation. The carbon-rich material contains large amounts of planetary-type rare gases; at high temperatures, these gases would have been lost. Further, at high temperatures (greater than 1400 K), reactions such as

$$C(s) + MgFeSiO_4(s) = MgSiO_3(s) + Fe(s) + CO(g) \qquad \text{(V-2)}$$

proceed strongly to the right, provided that gases are able to escape from the magma chamber often enough to keep the CO pressure less than 20 atm. In fact, the limited thicknesses of the reduced (low-iron) reaction zones found adjacent to cracks in olivine grains are consistent with this reaction having been limited by falling temperatures and/or inability of CO to escape following the later introduction of carbon.

The simplest model calls for introduction of the carbon-rich material by an impact-produced shock event and simultaneous shock production of the diamonds from graphite. How the carbon-rich material originated and how it was able to penetrate to the depths of more than ~1 km needed to produce olivine cumulates are unresolved questions. Possibly the rock overlying the olivine was removed in the same impact or by an earlier one. Because planetary-type rare gases are always present in chondritic material, there is reason to believe that the carbon-rich material was formed in the solar nebula; it would be most useful to find evidence that would confirm this, thus expanding the spectrum of known nebular materials.

The Aubrites

The aubrites consist almost exclusively of enstatite, $MgSiO_3$; the aubrite FeO/(FeO + MgO) ratios are extremely low, typically about 0.02 mol%. The group is sometimes designated the enstatite achondrites. The enstatite crystals are commonly large, achieving lengths as great as 10 cm. The rarity of plagioclase and the large size of the crystals implies an origin as a refractory cumulate, but other evidence shows that this simple picture is not complete.

Several aubrites are regolithic breccias containing large amounts of solar-type rare gases; another (Cumberland Falls) contains large inclusions of a (probably LL) chondrite. Thus, these meteorites were on the surface of a parent body for an extensive period, a fact that conflicts with an origin as a cumulate layer deep within a planet.

The aubrites show close affinities to the EL chondrites. Each aubrite contains at least 10 mg/g of Fe–Ni metal. In most cases, metallic silicon is dissolved in these Fe–Ni grains, the highest values corresponding to the value of 12 mg/g found in the EL chondrites. Silicate FeO/(FeO + MgO) ratios in EL chondrites and aubrites are essentially indistinguishable. One aubrite, Shallowater, contains 90 mg/g of Fe–Ni, and it and some others contain resolvable amounts of planetary-type rare gas.

These facts have led to the hypothesis that the aubrites did not originate by igneous processes but by mechanical fractionation processes occurring in the solar nebula. According to this scenario, their low metal contents are understood to represent a continuation of the metal–silicate fractionation trend that also produced the EH chondrites (high metal/silicate ratios) and EL chondrites (low metal/silicate ratios). The large pyroxene crystals are viewed as annealing (solid-state growth) effects produced by high temperatures, perhaps from a superluminous phase of the Sun; this view is consistent with the observation that the large pyroxene grains in Shallowater enclose small metal grains.

This nebular model does not offer a simple explanation of the low plagioclase content of the aubrites. The *ad hoc* proposal is that the grain size of the remaining nebular plagioclase (or its precursor phase) was too small to allow it to settle to the nebular midplane at the late time when the aubrites formed (see Chapter VII).

Returning to the igneous model, the absence of plagioclase can be explained by proposing that a basaltic layer that previously existed on the parent-body surface above the pyroxene cumulate had been removed by a massive cratering event. The cumulate was then exposed to the usual set of regolithic effects, including irradiation by the solar wind. The origin of the metal trapped in the Shallowater pyroxene crystals is still not readily explained by this model. Both of these aubrite models are intriguing; it will be interesting to see which of them receives more support from future studies.

Suggested Reading

Broecker, W. S., and V. M. Oversby. 1971. *Chemical Equilibria in the Earth.* McGraw-Hill. Chapter 1 gives a useful discussion of the advantages and limits of applying thermodynamics to earth science problems.

Consolmagno, G. J., and M. J. Drake. 1977. Composition and evolution of the eucrite parent body: evidence from rare earth elements. *Geochim. Cosmochim. Acta* 41:1271. Technical discussion of the rare-earth and petrological evidence bearing on the origin of the eucrites.

Dodd, R. T. 1981. *Meteorites: A Petrologic-Chemical Synthesis.* Cambridge University Press. Chapters 8 and 9 provide many details about the properties of the differentiated silicate-rich meteorites.

McCarthy, T. W., L. H. Ahrens, and A. J. Erlank. 1972. Further evidence in support of the mixing model for howardite origin. *Earth Planet. Sci. Lett.* 15:86. Discussion of compositional evidence in favor of howardites having formed by mixture of eucrite and diogenite components.

Stolper, E. 1977. Experimental petrology of eucrite meteorites. *Geochim. Cosmochim. Acta* 41:587. A technical discussion of the origin of the eucrites, based on experimental petrological studies.

Wasson, J. T., and C. M. Wai. 1970. Composition of the metal, schreibersite and perryite of enstatite achondrites and the origin of enstatite chondrites and achondrites. *Geochim. Cosmochim. Acta* 34:169. Technical but somewhat dated discussion of the evidence linking the aubrites to the EL chondrites and favoring nebular formation of the aubrites.

CHAPTER VI

Setting the Scene: Models of Solar-System Formation

There is no reason to think that the processes that led to the formation of the solar system were significantly different from those currently occurring in regions where new stars are being formed. Thus, key insights into the formation of the solar nebula can be obtained by the telescopic study of those regions where star formation is observed to be in progress. The amount of information that can be obtained is inversely related to the distance from the Earth, so the most important regions are those in our part of our galaxy. Our galaxy is called the Milky Way galaxy, a name derived from the bright band that on dark nights can be seen crossing the sky. The brightness of the band results from the fact that in this direction we are seeing the galaxy edge-on, greatly increasing the number of stars in our line of sight.

Star Classification and Evolution

Galaxies show a wide range of morphologies. The Milky Way galaxy is a spiral galaxy, a disk-shaped distribution of stars. Most of the stars visible from Earth are concentrated in "arms" that radiate spirally outward from the galactic center. The radius of the Milky Way galaxy is about 3×10^9 AU. The Sun is about 2×10^9 AU from the center, near the "inner" edge of an arm.

The mass of the Milky Way galaxy is about $2 \times 10^{11} M_\odot$ (where $M_\odot$ is the mass of the Sun). About 90 percent of the mass is in the form of $\sim 2 \times 10^{11}$ stars, the remainder in the form of gas and dust in the region between the stars, the **interstellar medium.**

As discussed in Chapter III, the age of the solar system is about 4.53 Gyr; it follows that the galaxy is older. The age of the galaxy can be estimated from astronomical observations combined with star-formation models and from the relative abundances of long-lived radioisotopes such as ^{232}Th, ^{235}U, and ^{238}U; it is about 10 to 18 Gyr old, with 13 Gyr its most-probable age.

The chief source of the energy required to make stars shine is the **nuclear fusion**[1] of hydrogen and helium. From the mass of the Sun and the rate at

---

[1] In **nuclear fusion,** two or more light nuclei react to form a heavier nucleus with the release of energy. By far the largest amount of energy release per gram of fuel occurs in the process of **hydrogen burning,** in which four nuclei of hydrogen are

which it emits energy, we can estimate that the Sun will use up its hydrogen about 5 Gyr from now. Thus the Sun is now halfway through its anticipated 10^{10}-yr life. The brightest stars are emitting light at rates about 10^6 times more rapidly than the Sun, but their masses are only about $100M_\odot$. Thus, their total supply of hydrogen fuel will be expended within about 10^6 years. It follows that these stars are far younger than the galaxy, and that star formation must be occurring in the galaxy today.

Young stars are commonly found in groups, often associated with dusty regions of space. Figure VI-1 is a negative image (the stars are black dots) of the Pleiades (also called the Seven Sisters), a cluster of stars of which at least six are easily visible with the naked eyes in the fall and winter sky from northern-hemisphere locations. The bright patches near the stars are remnants of the placental **interstellar cloud** from which the stars formed; their brightness results from the reflection by dust of light from the stars in the cluster. It is evident from Figure VI-1 that the whole region near the cluster has a higher-than-average dust density, because "background" stars of a given brightness are less numerous near the Pleiades cluster than they are near the edge of the photo.

Stars form in the interstellar medium when gravitational instabilities lead to contraction. Whether or not a particular region can undergo gravitational collapse depends primarily on the temperature T (in K), the concentration of matter (expressed as n_{H_2}, the number of H_2 molecules per cubic centimeter), and the total mass M_J of the region undergoing collapse (in multiples of $M_\odot$). This relation is commonly called the Jeans' criterion, and it is written in a way that emphasizes the competition between (1) thermally induced motions that tend to expand the cloud and (2) the increase in gravitational potential with increasing density that tends to cause contraction:

$$M_J \sim 20T^{3/2}n_{H_2}^{-1/2} \qquad\qquad \text{(VI-1)}$$

For example, a region in which $T = 10$ K and $n_{H_2} = 10^5\,\text{cm}^{-3}$ could undergo collapse if its mass were $2M_\odot$. In fact, this great a density is found only in the

---

fused to yield one helium nucleus. The net reaction is

$$4\,H^+ = {}^4He^{2+} + 2\,e^+ + 2\,\nu$$

where e^+ represents a positive electron and ν represents a neutrino. The reaction in fact occurs by two different mechanisms; each mechanism requires three or four steps.

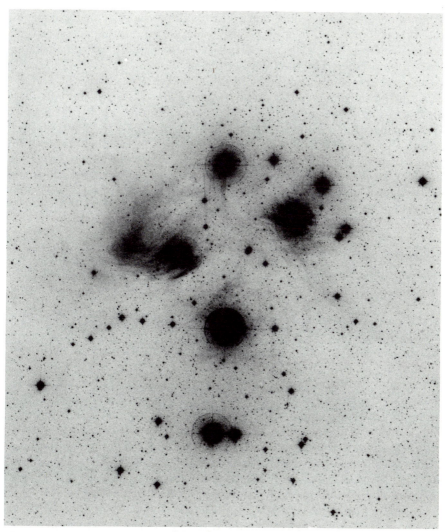

Figure VI-1. A negative print of the Pleiades, a star cluster now about 60 Myr old. Some of the original cloud from which the cluster formed is still visible in light reflected from dust in the cloud. It was necessary to overexpose the stars in order to reveal the distribution of dust. The halos and crosses associated with the star images are telescopic artifacts. (Print provided by G. O. Abell.)

cores of dense interstellar clouds; the more typical n_{H_2} value of $\sim 10^3$ cm^{-3} found in the envelopes of dense clouds leads to critical masses of $\sim 20 M_\odot$. These dense clouds are often called molecular clouds because of radio observations of formaldehyde (HCOOH), carbon monoxide (CO), and other molecular species. Rotation of the cloud produces centrifugal forces resisting collapse, and magnetic fields generate an internal pressure that hinders collapse. Therefore, actual critical masses are significantly greater than those predicted by the Jeans' criterion.

Stars are commonly classified into color–temperature classes O, B, A, F, G, K, and M (the mnemonic phrase is "*Oh Be A Fine Guy, Kiss Me*") in order of decreasing temperature and increasing redness. Figure VI-2 is a Hertzsprung–Russell diagram in which the luminosity of the star is plotted against the color or temperature; the class boundaries are shown at the top of the diagram. On such a diagram, most stars fall in a diagonal band called the **main sequence.** Theoretical calculations show that the stars on the main sequence are in the hydrogen-burning phase. The Sun is located near the $1M_\odot$ point on Figure VI-2; it is a typical G-type main-sequence star, a common type of star having below-average luminosity.

As stars evolve, they successively produce and then burn heavier elements. The rate at which the nuclear fuel is consumed is a function of its concentration, the density, and the temperature. The minimum temperature necessary to initiate the fusion of hydrogen to form helium is about 5×10^6 K. As hydrogen burning decreases the concentration of hydrogen in the core, the star compensates by contracting, thus raising both the central density and the temperature. Eventually, however, the hydrogen fuel in the core is reduced to such a low concentration that hydrogen burning stops.

Further contraction then occurs, and hydrogen burning becomes possible in a shell surrounding the hydrogen-depleted core. Continuation of this process increases the fraction of the star's mass that is depleted in hydrogen. Theoretical calculations show that an instability occurs when the depleted core reaches a critical size amounting to about 10 to 15 percent of the mass of the star. When this size is reached, the pressure in the core cannot support the overlying mass and, as a result, the core collapses and quickly reaches temperatures of $\sim 10^8$ K such that ^{4}He can "burn" to form ^{12}C. This nuclear reaction yields a low heat output per unit mass, and the collapse continues until pressures and temperatures are high enough to yield a high rate of ^{4}He burning. Simultaneously, the outer envelope of the star undergoes a major expansion; in the more massive stars, the radius may reach 1 AU. The luminosity of the star remains roughly constant during expansion of the envelope; thus the blackbody temperature decreases, and the color reddens. The resulting star is called a **red giant.** The position of red giants in the H–R diagram (Figure VI-2) is to the right of the middle and upper part of the main sequence.

The evolution of stars following the hydrogen-burning stage depends on the size of the star and is poorly understood. For medium-size stars, the sequence appears to include a red-giant stage and then a **planetary-nebula**

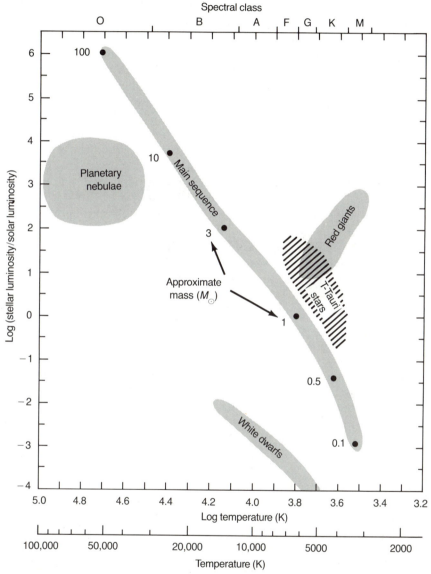

Figure VI-2. A Hertzsprung–Russell (H–R) diagram, in which luminosity is
plotted against temperature. On such a diagram, stars that are burning hydrogen
(the great majority) are distributed along a diagonal curve called the main
sequence. After using up their hydrogen fuel, stars evolve successively into he-
lium-burning red giants, carbon-burning planetary nebulae, and finally white
dwarfs that are cooling because their nuclear fuel has been expended. The
T-Tauri stars are young objects that have not yet evolved onto the main sequence.

stage (see Figure VI-2) in which a shell of gas is blown off, then heated by the bright residual star (possibly during the period in which it is burning ^{12}C to produce ^{24}Mg). Eventually the star reaches a stage where contraction does not produce further nuclear reactions in the interior, and it slowly cools, first to produce a **white dwarf** (see Figure VI-2), and finally to become so cool that it ceases to be telescopically visible. The density of white dwarfs is extremely high, $\sim 10^6$ g $\cdot$ cm^{-3}. White dwarfs with masses of $3M_\odot$ or less eventually radiate away their energy and become black dwarfs. More massive stars that have burned their nuclear field can contract to densities so great that light cannot escape the gravitational field. These invisible, superdense objects are called **black holes.**

The shaded fields are shown only for the well-populated regions of the Hertzsprung–Russell diagram in Figure VI-2; stars are found in other regions as well. The general reason for the lower populations in these other regions is that typical stars evolve through them rapidly. As an example, main-sequence stars having masses greater than $5M_\odot$ evolve roughly horizontally into red giants as less-massive stars do, but the evolution is at such a rapid rate that few of these supergiants appear in star surveys. Similarly, the evolution from the planetary-nebula stage into white dwarfs is initially rapid but slows down appreciably as the star's surface area is reduced by contraction.

The final evolutionary stage of the largest stars is a massive explosion of a red giant to form a supernova. Figure VI-3 shows a photograph of the Crab Nebula, the remnants of a supernova that exploded in our galaxy. The light from the explosion reached the Earth in the year 1054 and was observed by Chinese astronomers. During its period of maximum brightness, the luminosity of a supernova may exceed that of an entire galaxy. The remnant cores of supernovae are either superdense **neutron stars** or black holes.

There appears to be a consensus for the following scenario describing the source of the enormous amounts of energy liberated by supernovae. A massive star burns up all its nuclear fuel in the central regions; the hydrogen is converted to ^{56}Fe, the most stable nuclear species, through the intermediate forms of ^{4}He, ^{12}C, ^{24}Mg, and so on. Contraction eventually produces such high temperatures in the core that individual thermal photons have energies greater than 10 MeV and are capable of producing nuclear reactions. These reactions *absorb* energy — leading to further contraction and still-higher temperatures and thus to an acceleration in the rate of energy absorption. The result is an explosive collapse of the stellar core. The overlying shells participate in the collapse, but they still have residual fuel, ^{12}C in particular. Their inward collapse causes rapid increases in their pressure and temperature and an explosive acceleration in their rate of ^{12}C burning. The shells release energy far more rapidly than it can be transported to the

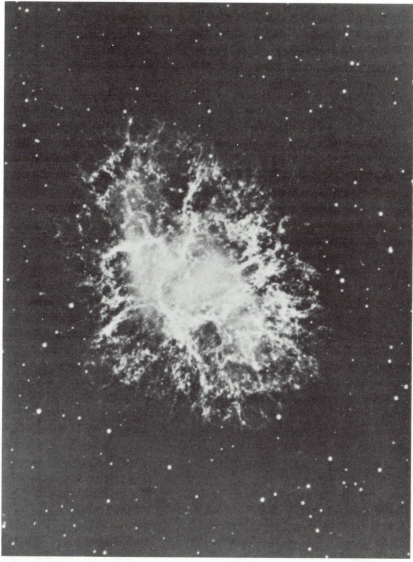

Figure VI-3. The Crab Nebula in the constellation Taurus consists of the remnants of a supernova that exploded in 1054 A.D. The irregular pattern of luminosity results from turbulent interaction between the high-velocity ejecta and the gas of the surrunding interstellar medium. Supernovae are believed to constitute the major source of elements heavier than iron in the interstellar medium. (Lick Observatory photo.)

surface, even by convection, and so an explosion occurs that removes the outer layers of the star.

The materials removed by the explosion expand rapidly into the surrounding medium. From the Doppler shifts in the wavelengths, the current expansion velocity of the Crab Nebula is estimated to be ~ 1450 km $\cdot$ s^{-1}. The radius is estimated to be $\sim 1.8 \times 10^5$ AU ($= 5.3 \times 10^{13}$ km), and formation in A.D. 1054 (2.9×10^{10} s ago) implies an expansion velocity of ~ 1800 km $\cdot$ s^{-1}, similar to that estimated from the wavelength shift.

After the explosion, the supernova's luminosity decreases rapidly, initially with a half-life of about 50 days. Eventually, the residual central star becomes a minor contributor to the brightness, as is the case for the Crab Nebula (Figure VI-3). The present brightness of the Crab Nebula results from the collision of the supernova ejecta with interstellar matter. The filamentary structure is caused by turbulence in the supernova-ejecta–interstellar-matter system.

Initiation of the Star-Formation Process

Many dense interstellar clouds have combinations of masses, densities, and temperatures such that the Jeans' criterion indicates that they should be undergoing gravitational collapse. However, if they are all collapsing, the rate of formation of new stars should be 30 to 100 times higher greater than that observed. These clouds must be stabilized by effects not included in the Jeans' criterion, such as magnetic forces and angular momentum, that prevent their collapse. It now appears that most clouds are stable until collapse is triggered by an external influence.

Two collapse triggers have been widely discussed during recent years: (1) galaxy-wide density waves, and (2) shock effects produced by expanding supernovae remnants. It is now believed that the arms of spiral galaxies such as the Milky Way galaxy are so dramatically visible not because they have a higher density of matter than the interarm regions, but because they are the regions where star formation is occurring and as a result contain a large fraction of the young, massive stars that provide most of the luminosity.

The stars and the interstellar matter move in orbits about the center of the galaxy. In our part of the galaxy, these orbits have periods of a few hundred megayears. In the density-wave model, the positions of the spiral arms are viewed as standing waves. When the galactic matter rotates into these regions, it undergoes compression, thus triggering star formation in dense clouds that were stable in the interarm region. Continued rotation brings the material out of the arm region and produces a rarefaction that stabilizes those clouds that had evolved to the brink of collapse. If there are two spiral arms in the galaxy, material near the Sun passes through an arm about once each 10^8 yr.

The collision of high-speed debris from a supernova with a dense interstellar cloud should cause compressions capable of triggering star formation. The signature of such a phenomenon would be a circular arc of star formation on the edge of a cloud. Astronomical observations of such features are now known.

Figure VI-4 shows a region called the Canis Major R1 association that

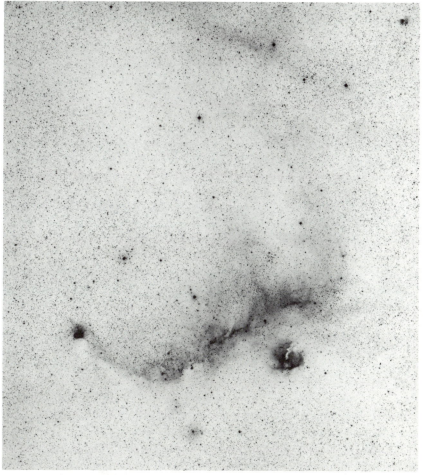

Figure VI-4. This negative print shows a cloud of luminosity in the constellation Canis Major. The generally circular shape of the luminous gas is attributed to excitation by the spherically expanding shell of a supernova that exploded about 1 Myr ago. Numerous young stars are found where the spherical shell collides with a dense cloud near the bottom of the photo. The presence of the dark cloud is recognized by the lower density of background stars. The younger stars are below the dark arc and are marked by halos of luminous gas. Width of photo $\sim 7 \times 10^6$ AU. (Mt. Wilson Observatory photo, provided by G. J. Wasserburg.)

contains many young stars. The R1 association is a group of new stars forming a sharp arc on the middle and lower-right part of the photo. The shape of the right edge of the association is generally circular, and the radius of curvature appears to be about one-half of the distance between the star-rich region on the lower right and that on the upper left of the figure. Much of the light in these regions is being emitted by the gas; this luminosity probably reflects the interaction of interstellar matter with turbulent debris from a supernova at a stage about 10^6 yr later than that of the Crab Nebula. Such emission rings are relatively common.

Most of the young stars lie just at the interface between the emission ring and a dark (but, for the most part, not completely opaque) cloud. This cloud is believed to define the inner edge of the supernova shock front. The youngest stars (commonly surrounded by clouds of luminous gas) lie to the right of the interface (away from the supernova locus), indicating that the shock front has just initiated star formation at this location.

Many of the stars in the Canis Major R1 association lie just above the main sequence, as expected of young stars evolving toward the main sequence. A few of the largest stars have evolved onto the main sequence. Theoretical models indicate that the smallest of these needed 3×10^5 yr to evolve to its present position. This young age is generally consistent with the 1-Myr age of the supernova calculated from its *size* (radius $\sim 6.5 \times 10^6$ AU $\cong 1 \times 10^{15}$ km) and an estimated mean expansion velocity of ~ 30 km $\cdot$ s^{-1}. The much lower expansion velocity relative to the Crab Nebula reflects the gradual deceleration of the ejecta by momentum exchange with the interstellar medium.

Observations of Stars Evolving Toward the Main Sequence

The earlier a star's position on its evolutionary track toward the main sequence, the more difficult it is to observe. At first it is subluminous, and then it passes through a highly luminous stage in such a brief period that the chance of observing it is small. Further, during either of these early stages, the star is likely to be obscured from our view by its placental dust cloud.

Figure VI-5 is a Hertzsprung–Russell diagram showing calculated pre–main-sequence evolutionary paths. The calculations are based on spherical symmetry and include no allowance for angular momentum or magnetic effects. Although this simplified model gives unrealistically short times, this discrepancy is probably partially counteracted by the fact that significantly more mass was initially involved than the mass finally incorporated in the star. Pre–main-sequence periods vary from $\sim 10^4$ yr for $100 M_\odot$ stars to about 10^9 yr for $0.1 M_\odot$ stars.

These calculations suggest that a star of mass about $0.5 M_\odot$ to $1 M_\odot$ would

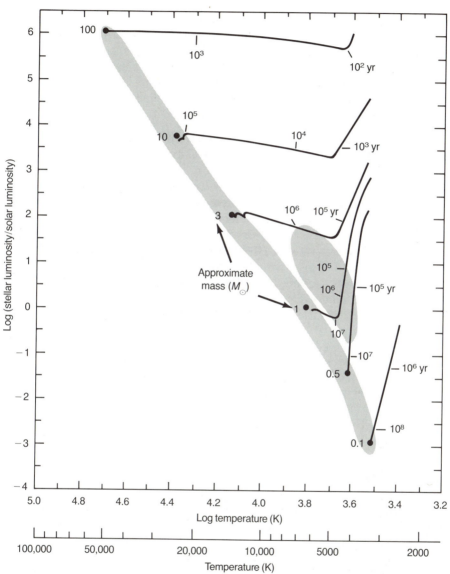

Figure VI-5. A Hertzsprung–Russell diagram showing the main sequence (diagonal band), the T-Tauri field (stippled area above main sequence), and calculated evolutionary paths leading to the main sequence. During the latter part of the evolutionary path, stars in the mass range of $0.5M_\odot$ to $1.0M_\odot$ can have luminosities up to 1000 times greater than their main sequence value. See text and Figure VI-2 for more details. (Evolutionary paths from P. Bodenheimer, *Rep. Prog. Phys.* 35:1, 1972.)

be 1000 times more luminous early in its evolution than it will be after arriving on the main sequence. This is the superluminous period discussed as a possible parent-body heat source in Chapter IV. Such a superluminous period would last only about 10^3 yr, and it may not have existed at all if angular momentum and magnetic effects braked the rate of evolution.

A combination of observational and theoretical efforts has provided various ways to recognize regions where star formation is occurring.

1. As mentioned earlier, an O-type star has a lifetime of about 10^6 yr; thus any stellar cluster containing several O-type or (slightly less massive) B-type stars is probably young.

2. T-Tauri stars are characterized by irregular variability and Doppler-shifted emisson-line spectra. The emission lines originate in a nebula surrounding the star, and the fact that they are Doppler-shifted indicates that the star is losing (or, possibly in some cases, gaining) mass at remarkably high rates of $10^{-6}M_\odot$ to $10^{-7}M_\odot$ per year, about 10^7 to 10^8 times the rate of solar-wind loss from the sun. Most T-Tauri stars fall in the luminosity – temperature field shown in Figure VI-5 (and, for comparison with more evolved stars, in Figure VI-2). Their ages, as inferred from the ages of the clusters they are found in, are consistent with the ages of 10^5 to 10^7 yr implied by the theoretical evolutionary paths.

3. Because infrared light is less-strongly absorbed by dust than is visible light, infrared observations can reveal hot objects hidden in dark clouds. A few highly luminous objects having inferred ages of $\sim 10^5$ yr have been discovered by this technique. It is not possible to distinguish whether (1) the energy is being liberated entirely by a hot central star and then converted to infrared by scattering from dust particles, or (2) some or most energy is being produced directly in the stellar nebula as a result of gravitational collapse. The later possibility is particularly intriguing.

A nearby region ($\sim 10^8$ AU away) where star formation is occurring is the Orion Nebula, centered in the sword of the Orion constellation; Figure VI-6 shows this handsome gas – dust – star association. Within the nebula are found each of the features just mentioned: associations of O and B stars, associations of T-Tauri stars, and strong infrared sources embedded in obscuring dust.

A generally circular shape of the lower-right edge of the nebula is easy to imagine. Barely visible in the upper-left portion of Figure VI-6 are luminous emission features that seem to be segments of the same circle. This loop, with a radius of about 8×10^6 AU, is suggestive of a supernova explosion. Calculations indicate that several large supernovae are required to generate the observed mass motions, but considerations we will discuss shortly indicate that this is not an implausible scenario.

Figure VI-6. The Orion Nebula is the large cloud in the upper-central part of this telescopic photo. Star formation is now active in a region called the Trapezium, just to the left of the dark embayment that roughly bisects the large bright arc in the lower-right part of the nebula. The generally circular shape of the bright emission regions may have resulted from several supernova explosions. (Lick Observatory photo.)

Figure VI-7 shows the location within the Orion Nebula of a series of groups of stars of differing age. Stars of the oldest group are shown by the crosses, those of the second-oldest group by open circles, those of the third-oldest group by filled squares, and the several hundred stars belonging to the youngest group all fit inside the shaded circle. The ages of these groups are estimated to be 7.9, 5.1, 3.7, and less than 0.5 Myr, respectively. The roughly linear arrangement of the first three groups is consistent with successive triggering of star formation in four groups. Although the third and fourth groups are seen superposed by an observer on Earth, the youngest group (named the Trapezium, after four very bright O and B stars that form a trapezoid) lies farther away from us than the third group, and its location is consistent with triggering by a supernova from the third group.

Note also that Figure VI-7 suggests that the linear dimensions of a stellar association increase with increasing age; this appears to be a generally valid observation. Apparently, most stars form with randomly oriented velocities relative to the center of the cluster of ~ 3 km $\cdot$ s^{-1}, sufficient to escape the gravitational field of the cluster. If the Sun left its cluster with such a velocity 4.6 Gyr ago, it has now traversed $\sim 2 \times 10^{10}$ AU relative to any

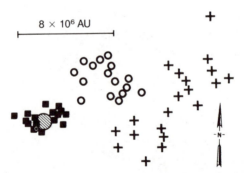

8×10^6 AU

Figure VI-7. Star clusters are commonly recognized from a combination of spatial nearness and similar motions among their members. Relative locations of the members of four such clusters recognized in the lower-right portion of the Orion Nebula are illustrated here. Members of the oldest cluster are represented by crosses, those of the next-oldest by open circles, and those of the third-oldest by filled squares. The numerous members of the youngest cluster (including the Trapezium stars) are inside the shaded circle. The older the cluster, the greater the mean separation of its members. Ages estimated from the evolutionary status of massive stars range from 8 Myr in the oldest group to less than 0.5 Myr in the youngest. It has been suggested that supernova explosions in the older groups served to trigger star formation in the next younger groups. (After a drawing by H. Reeves in *Protostars and Planets*, ed. T. Gehrels, University of Arizona Press, 1978, p. 399.)

residual cluster, far too great a distance to allow us hope of recognizing our birthplace.

Stars that escape the cluster also escape the interstellar cloud that gave them birth. In the case of clusters like the Pleiades (Figure VI-1), the stars and the residual cloud clearly can remain in proximity for as much as 10^8 yr, though it seems certain that extensive dissipation of the cloud at the Pleiades location has already occurred, and thus that 100 Myr is probably near the maximum period of star – cloud association.

Formation of the Protosun and the Solar Nebula

Although the last few years have yielded significant progress in understanding the conditions and processes leading to star formation, we are still far from a detailed picture. As a result, the following scenario is only a very rough outline of the events that produced the early solar system.

The galaxy formed some 10 to 18 Gyr ago as part of a general period of galaxy formation. Star formation occurred by processes similar to those discussed in the preceding section. Some short-lived stars ejected a portion of their processed interior materials, including the heavy elements, back into the interstellar medium. Supernovae were probably the dominant sources of these heavy elements.

About 4.5 Gyr ago, the Sun formed in a cloud of unknown size. As discussed in Chapter III, we see evidence of ^{129}I formation in an event ~ 100 Myr before the solar system formed, perhaps in a star formed during the previous passage of local material (I avoid using "local cloud" because, as indicated earlier, distinct clouds probably dissipate within ~ 100 Myr) through a density wave.

If the ^{26}Al that produced the observed ^{26}Mg anomalies existed when the meteorites formed, then the Sun could not have been among the first stars to form in the cloud. Rather, there was time enough for at least one of the more massive stars to evolve into a supernova and eject fresh ^{26}Al and the other isotopes into the region that spawned the Sun.

The protosun formed by collapse of a region having a mass of at least $1.01 M_\odot$, because $0.01 M_\odot$ is the estimated minimum amount of mass that can provide the nonvolatile fraction of the planets. Some models call for $1 M_\odot$ in the planetary region, but there appears to be a modest consensus favoring the view that there was less than $0.1 M_\odot$ in this region. The problems with assuming small amounts of mass in the planetary region are (1) how to heat it to the temperatures inferred fron the properties of chondrites (see Chapter VII) and (2) how to form Uranus and Neptune on a time scale of $\sim 10^9$ yr. The problems with assuming a large nebula are (1) how to dissipate and remove the unneeded material, and (2) how to prevent the formation and survival of giant planets in the inner solar system.

The release of gravitational energy brought the nebula to high temperatures (1000 K, and possibly higher). Because of the high temperatures and the high velocities of the collapsing matter, the nebula was turbulent and thus well stirred. It seems unlikely that isotopic inhomogeneities in the gas phase could survive this stage.

Eventually, the turbulence damped out, and the extrasolar material collapsed down to a disk-shaped cloud surrounding the Sun. It seems probable that the nebula continued to accrete matter, and that the resultant energy release produced an extended period of moderately high temperatures. A refractory portion of this late-accreted material could have survived and thus preserved isotopic anomalies.

The preceding scenario seems an adequate introduction to the nebular processes that left a recognizable record in the chondrites. Other models for the introduction of the extrasolar matter into the solar nebula (such as tidal disruption by a passing star, or slow capture of material from a tenuous plasma) seem unnecessarily complex, and they currently have little support among researchers in the field. Some complexities may be unavoidable, however—such as initial formation of the Sun as part of a multiple-star system in order to account for the 7° difference between the spin axis of the Sun and the axis of rotation of the planets about the Sun.

Suggested Reading

Andouze, J., and S. Vauclair. 1980. *Introduction to Nuclear Astrophysics.* Reidel. A moderately technical but fluent treatment of star evolution and the stellar synthesis of their elements.

Cameron, A. G. W. 1975. The origin and evolution of the solar system. *Sci. Amer.* 233(3):32. (Reprinted in *The Solar System,* Freeman 1975.) Nontechnical discussion of star formation and the formation of planets.

Herbst, W., and G. E. Assousa. 1978. The role of supernovae in star formation and spiral structure. In *Protostars and Planets,* ed. T. Gehrels, pp. 368–383. University of Arizona Press. Moderately technical discussion of the relative roles of supernovae and galactic density waves in triggering star formation.

Meadows, A. J. 1978. *Stellar Evolution.* 2nd ed Pergamon. A nontechnical discussion of star formation and evolution.

Reeves, H. 1978. The "bingbang" theory of the origin of the solar system. In *Protostars and Planets,* ed. T. Gehrels, pp. 399–426. University of Arizona Press. A relatively technical discussion of the role of supernova explosions in triggering star formation.

Strom, S. E. 1976. Star formation and the early phases of stellar evolution. In *Frontiers of Astrophysics*, ed. E. Avrett, pp. 95–117. Harvard University Press. A nontechnical description of early stellar evolution.

Tayler, R. J. 1970. *The Stars: Their Structure and Evolution*. Wykeham. A moderately technical description of star formation and evolution.

Wood, J. A. 1979. *The Solar System*. Prentice-Hall. Chapter 6 describes star formation and evolution in nontechnical fashion.

CHAPTER VII

Chondritic Meteorites as Products of the Solar Nebula

The chondritic meteorites have solar compositions that are consistent with a primitive origin (formation in the solar nebula) and inconsistent with their having experienced melting and igneous fractionation. As discussed in Chapter III, these meteorites are extremely old; all contain excess ^{129}Xe, evidence of formation within about 100 Myr after the synthesis of ^{129}I in a supernova. Some of their parts contain excess ^{26}Mg, evidence of formation within a few megayears of the time of the last presolar supernova explosion.

As discussed in Chapter VIII, some current star-formation models do not yield high temperatures in the solar nebula. If temperatures were never high enough to vaporize the presolar grains, then chondritic meteorites should consist of interstellar grains. Conversely, if the chondrites show properties that require high nebular temperatures, then there is an implication that something is wrong with the star-formation computations and that one should consider changing some of the assumptions that enter those calculations.

In this chapter, I discuss (1) the evidence indicating that chondrites are not simply collections of interstellar grains, and (2) the evidence that constrains the maximum temperatures reached by chondritic materials.

A unique feature of the chondrites is the presence of chondrules, millimeter-size spheroidal objects that formed as partly or wholly molten droplets. Much effort has gone into attempts to understand the origin of these enigmatic grains, with a mixed degree of success. In the final portion of this chapter, evidence obtained from the study of chondrules is summarized and used to evaluate formation models.

Evidence Against an Interstellar Origin for Most Chondritic Grains

There is strong evidence that the chondrites originated in the solar system and that they are not simply collections of interstellar grains. For the chondrite classes other than CM and CI, formation in the solar system is indicated by the observed textural and compositional properties and by the intergroup variation in these properties. The properties that vary between groups (such as refractory abundances, iron abundance, degree of oxidation, and chondrule size) are summarized in Chapter II. In order to account for these variations with a model of interstellar origin, one would have to hypothesize many different reservoirs of interstellar grains differing in size and composition, with a few of these reservoirs providing the bulk of the material to each group. Such a segregation of materials seems inconsistent with the large amount of turbulence and mixing expected during the decrease in linear dimension by a factor of $\sim 10^4$ needed to produce the solar nebula from an interstellar-cloud fragment. The general constancy of isotopic ratios in elements other than oxygen seems to imply a thorough mixing of most preexisting reservoirs rather than their segregation. Furthermore, if chondrites formed from interstellar grains, irradiation of these grains by cosmic rays should have produced large isotopic effects that are not observed.

It is not so certain that large fractions of the CI chondrites or of the fine-grained matrix materials of CM chondrites are not of interstellar origin, because the "solar" compositions of these materials are near that expected for a random collection of interstellar materials, and the individual minerals are highly unequilibrated. However, the weight of the evidence favors the conclusion that the low-temperature components of these meteorites formed by condensation or by aqueous alteration of preexisting materials. The high-temperature grains found in the CM chondrites do not contain the amounts of cosmogenic rare-gas isotopes expected in unaltered interstellar materials. Either these grains are interstellar and lost their presolar rare gases during a heating event that altered but did not vaporize the grains, or they formed mainly by condensation or melting processes in the solar nebula.

In most of the following discussion, the following working model of the solar nebula is assumed.

1. Its mass was the minimum necessary to account for the masses of the planets, about $0.01M_{\odot}$ to $0.03M_{\odot}$; although models involving more massive nebulas have been discussed, no one has proposed a convincing physical mechanism for removing the excess material from the solar system.
2. It formed relatively rapidly and thus experienced appreciable heating by the release of gravitational energy.
3. Its outer regions were opaque to infrared radiation, and thus it cooled relatively slowly.

Evidence that Chondrites Formed in the Solar Nebula

Several features of chondritic meteorites attest to their formation in the solar nebula rather than on or in parent bodies. Their compositions are solar; abundances of all but the most volatile elements are within a factor of 2 of solar values (see Chapter II). These compositions are inconsistent with a magmatic origin; igneous rocks have narrow melting ranges of ~50 K, whereas chondrites have much larger melting ranges of ~200 K. The unequilibrated chondrites preserve to a greater or lesser extent a disequilibrium state that is inconsistent with a magmatic origin; for example, adjacent olivine grains often differ greatly in their FeO/(FeO + MgO) ratios, with values ranging from less than 0.01 to ~0.5. The grains in these unequilibrated (petrologic types 1 through 3, see Chapter II) chondrites also preserve oxygen isotopic differences that are inconsistent with the isotopes ever having been completely equilibrated. The unequilibrated chondrites contain large quantities of planetary-type rare gas that was almost certainly incorporated in nebular and, for some rare components, prenebular (interstellar) processes, although the exact nature of each set of processes remains obscure.

The ages of many chondrites are high, unresolvable from the 4.53 Gyr age (Chapter III) of the Solar System. Figure VII-1 is a histogram showing chondrite ^{39}Ar–^{40}Ar plateau ages (see definition in Chapter III) published through 1979. Many of these chondrites were selected for analysis because

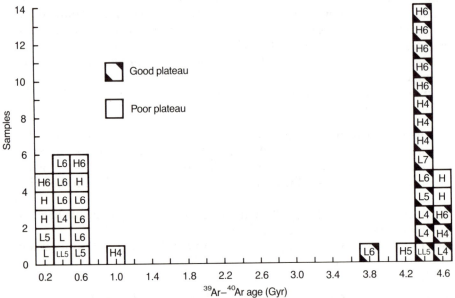

Figure VII-1. Plateau ^{39}Ar–^{40}Ar ages (see Chapter III) of ordinary chondrites. Most samples are heavily shocked, but only about one-half of them show ages measurably less than the 4.53 Gyr age of the solar system. (Data mainly from D. D. Bogard et al., *J. Geophys. Res.* 81:5664, 1976; G. Turner, *Meteorite Research*, p. 407, Reidel, 1969; and G. Turner et al., *Proc. Lunar Planet. Sci. Conf. 9th*, p. 989, 1979.)

they show evidence of heavy shock and were expected to have low plateau ages. Some do have ages of 0.1 to 0.7 Gyr, but a large peak at high ages also exists. Even those chondrites whose ages to a greater or lesser extent have been reset by shock events (such as those showing low $^{39}Ar - ^{40}Ar$ ages in Figures III-3 and VII-1) often retain the record of $^{129}I - ^{129}Xe$ formation intervals that are the same within 10 to 20 Myr and that testify to formation very early in solar-system history (see Figure III-4 and Table III-2). The narrow range of formation intervals recorded by the chondrites is further evidence against each group having formed from select reservoirs of interstellar grains, because the apparent formation ages of these reservoirs would be expected to show wide variations reflecting gross differences in time of formation and in the degree of retention of fossil ^{129}Xe.

The magnitude and uniformity of the formation ages of the chondrites are strong evidence for their formation (1) earlier than the planets, and (2) by processes that occurred nearly simultaneously (to within a few megayears) throughout the entire range of formation regions, probably throughout the inner solar system. If the heating events that reset the isotopic clocks occurred in or on large ($r \geq 100$ km) parent bodies, one would expect many of these ages to have recorded the early intense bombardment that reset the ages of most highlands rocks to 3.9 to 4.0 Gyr; in fact, the ages plotted in Figure VII-1 show no tendency to peak in this range.

There are several sources that could have been responsible for the early heating event recorded by the formation-age clocks; one is primary (the heating of the solar nebula), and the others are secondary (events occurring after the agglomeration of chondritic materials to form rocks). The next section summarizes evidence indicating that the nebula was initially hot enough to vaporize at least 95 percent of the precursor interstellar solids in the regions where most chondrites formed; the source of this heat was probably gravitational (accretional) energy released during the formation of the solar system from an interstellar cloud.

Because most age measurements have been carried out on meteorites showing some evidence of metamorphic recrystallization, it is also possible that this metamorphic reheating reset the isotopic clocks. The metamorphic reheating could have resulted from ^{26}Al decay, collisional heating, heating by a superluminous Sun, or electrical heating by planetary currents induced by a supersolar wind. Arguments given in Chapter IV indicate that these sources are at best marginally able to melt the parent planets of the differentiated meteorites, but only about one-third as much energy is required to produce the observed metamorphic effects, and one or more of the heat sources may well have been equal to this reduced task. The narrow range of $^{129}I - ^{129}Xe$ and $^{87}Rb - ^{87}Sr$ formation intervals recorded in Figures III-4 and Table III-2 requires that the heat source have a short duration (10 to 50 Myr), consistent with ^{26}Al or with the supersolar-wind-induction heating (Chapter IV). It is questionable, however, whether enough metamorphism occurred to reset the ^{129}I–^{129}Xe clocks in the least-equilibrated chondrites for which old formation intervals have been recorded (Allende and Chainpur samples in Figure III-4); thus, these ages almost certainly record formation in the nebula. The simplest and seemingly most plausible interpretation of chondrite formation ages near 4.5 Gyr is that they reflect a high degree of isotopic mixing in the solar nebula, followed by minimal amounts of isotopic redistribution during metamorphism.

As noted in Chater III, planetary-type rare gas (see Figure III-8) is found in all chondritic meteorites. It is not clear how it is incorporated, but the consensus seems to favor the hypothesis that it resulted from adsorption of the He, Ne, Ar, Kr, and Xe onto grain surfaces from a nebular mixture that had solar proportions of the elements. *Ad*sorption refers to the physical or chemical attachment of a gaseous atom to a host surface, whereas *ab*sorption refers to the incorporation of a substance throughout the volume of the host. It is well established that the degree of adsorption increases with increasing atomic number of the rare gas. Because no plausible planetary fractionation mechanisms have been proposed (despite the designation "planetary type"), the presence of this gas offers some support to the conclusion that chondrites are nebular products.

The preceding evidence indicates that most of the properties of the unequilibrated chondrites were established in nebular processes. Thus, the detailed properties of these chondrites (and, to some degree, those of their metamorphosed equivalents) can be used to infer the properties of the solar nebula. Because the ten groups as well as the ungrouped chondrites seem to have formed at different distances from the Sun, we can also examine the spatial variations in nebular conditions. As will be discussed in more detail later in this chapter, there is also reason to believe that sophisticated studies will eventually reveal a time sequence among the various grains (a rough sequence is already inferred from early, high-temperature phases through late, low-temperature phases), thus leading to information about the temporal variation in nebular properties.

Maximum Temperatures in the Solar Nebula

Several features of the chondrites appear to require high temperatures early in the history of the solar nebula. It is important to try to quantify this evidence because initial temperatures are an important parameter both (1) for models of the formation of the solar system from an interstellar cloud fragment and (2) for more-detailed models attempting to account for the properties of individual chondrites or chondrite groups.

In Chapter IX, it is argued that the chondrite groups formed over a large range of distances from the Sun, perhaps 0.4 to 4 AU. Despite origins at diverse locations, most of these groups are remarkably uniform in isotopic composition. If we ignore cosmogenic and radiogenic effects and the fractionations observed in the highly volatile rare gases, the only element known to vary in isotopic composition from group to group is oxygen. Investigations of the compositions of numerous elements (K, Mo, Cd, Ba, and several rare earths) have revealed no detectable variations among groups. This observation implies either the absence of major isotopic variations among the various components of interstellar matter or a thorough homogenization of these components during solar-system formation. Because a few chondritic components (particularly the refractory inclusions in CV chondrites) do show anomalous isotopic compositions (always found in oxygen and titanium, in a few cases in other elements), homogenization is the more plausible alternative. If the interstellar material consisted of dispersed fine grains ($r < \mu$m), a thorough mixing of these grains followed by mild metamorphism would produce homogenization. However, if some materials were coarse grains or grain assemblages (as required if the large refractory inclusions in CV chondrites are residues), then this coarse material must have been vaporized or crumbled before the formation of most meteorite groups. The arguments to follow suggest that vaporization was the dominant mechanism that produced this homogenization.

Several of the arguments used to estimate maximum nebular temperatures are based on comparison of chondrite mineralogy and composition with that expected from equilibrium condensation in a cooling nebula initially having a solar composition. The basis for such calculations is described in Appendix F. Table VII-1 lists 50%-condensation temperatures for highly

Table VII-1
The 50%-condensation temperatures (in K) under equilibrium conditions in a cooling solar nebula

Element	Condensate	Nebular pressure (atm)				
		10^{-6}	10^{-5}	10^{-4}	10^{-3}	10^{-2}
Mg	Mg_2SiO_4	1203	1268	1340	1417	1494
Ca	$Ca_2Al_2SiO_7$	1382	1447	1518	1596	1683
Ti	$CaTiO_3$	1420	1482	1549	1622	1703

Source: J. Wasson, in *Protostars and Planets*, ed. T. Gehrels, p. 488, University of Arizona Press, 1978.

refractory titanium, moderately refractory calcium, and the common element magnesium at a series of pressures spanning the plausible range for the solar nebula inside 4 AU.

Figure II-3 illustrates the parallel fractionations of refractory abundances among the chondritic groups except LL (which is essentially identical to L) and IAB, for which the data scatter but with typical values in the range between those in L and those in EH and EL chondrites. If the nebula was as well homogenized as the arguments given earlier indicate, and if the CI chondrites preserve mean solar abundances, then a refractory component is partially missing (~50 percent has been "lost") from the EH and EL chondrites and is present in ~35 percent excess in the CV chondrites. Petrographic examination of the enstatite and ordinary clans shows small-to-negligible amounts of refractory inclusions; the refractories revealed by chemical analysis are almost entirely present as unrecognized small grains or have been incorporated into the moderately sized (0.1 mm) grains formed by the common elements. In contrast, a substantial fraction of the refractories in the CV chondrites appear as millimeter- to centimeter-sized grains that are orders of magnitude larger than the typical grains that carry the common lithophiles magnesium and silicon (see Figures VIII-2 and VIII-4). It seems clear that the physical properties of both refractory and common components varied from location to location in the solar nebula. It is probable that this variation can be explained in terms of maximum temperatures, cooling rates, pressures, and so on that varied from location to location but, at most locations (perhaps all except the CV and CM–CO), initial temperatures high enough to vaporize nearly all (more than 95 percent) of the preexisting interstellar solids are implied. The data in Table VII-1 show that this hypothesis requires temperatures ranging from ~1450 K at 10^{-6} atm to ~1730 K at 10^{-2} atm.

At the CV location, we can be more specific about the maximum temperatures. As discussed in Chapter V, igneous fractionations of the rare earths other than europium produce only modest fractionations that are correlated with the systematic decrease in the $+3$ ionic radius from lanthanum through lutetium. The irregular rare-earth patterns determined in some refractory inclusions in CV chondrites (an example appears in Figure VII-2) cannot be explained by fractionations occurring in igneous processes. In contrast, differences in condensation behavior can lead to precisely the observed fractionation pattern. The calculated points in Figure VII-2 show the rare-earth spectrum expected if the most refractory rare earths condensed into $CaTiO_3$, which was withdrawn to an unknown location (probably into a kind of refractory material not yet studied) before the inclusion having the unusual pattern formed. This isolation of the $CaTiO_3$ must have occurred after $CaTiO_3$ had condensed but before most calcium had condensed as $Ca_2Al_2SiO_7$. (Because calcium is much more abundant than tita-

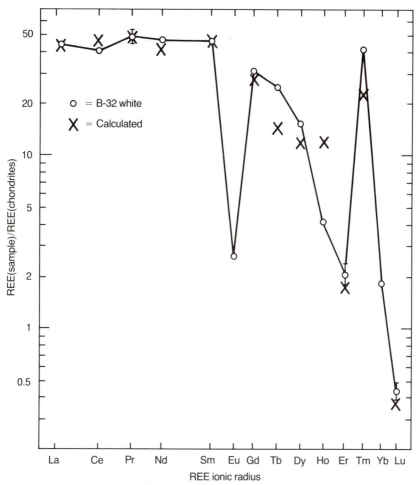

Figure VII-2. A highly irregular rare-earth pattern in the refractory inclusion B-32 white from the Allende CV chondrite cannot be explained by igneous fractionations, because these processes produce only small fractionations between adjacent rare earths having +3 oxidation states, whereas erbium and thulium are fractionated by a factor of more than 10. However, calculations by W. Boynton (*Geochim. Cosmochim. Acta* 40:63, 1978) show that the pattern is that expected in a refractory condensate that formed after perovskite ($CaTiO_3$) had condensed and been isolated from nebular gases.

nium, only 3.5 percent of nebular calcium can condense as $CaTiO_3$.) The CV chondrites probably formed at relatively low pressures far from the Sun; if the nebular pressure were 10^{-5} or 10^{-6} atm, the data in Table VII-1 show that maximum temperatures reached about 1460 or 1400 K, respectively.

In Chapters III and VIII, evidence is presented indicating that some of the refractory materials in CV chondrites are unevaporated residues of interstellar solids, implying that the maximum temperatures at the CV location did not exceed the condensation temperature of $CaTiO_3$ after the accretion of presolar matter to this nebular location was complete.

In this discussion, I have treated temperature as if it were constant at each moment in time for every radial distance from the Sun. In fact, there was a maximum temperature at the nebular median plane. The decrease in temperature with height perpendicular to the plane depended on the opacity and the rate of deposition of accretional energy with height. If there was no heat input and if the temperature decreased adiabatically with height, then about 75 percent of the gaseous mass of the nebula had temperatures at least 0.90 times as high as those at the median plane. Because (as discussed in more detail in Chapter VIII) agglomeration of grains to form rocks is believed to occur in the nebula plane, a uniform temperature at each location and time is probably a rather good approximation.

Chondrules: Their Properties and Proposed Origins

Chondrules are millimeter-sized objects found in all chondrite groups except CI. Because of the substantial body of evidence indicating that unequilibrated chondrites experienced little alteration following formation in the solar nebula, it follows that chondrules also formed by nebular processes.

Before discussing the properties of chondrules, it is useful to define **matrix** to mean interchondrular material. Some fraction of this interchondrular material consists of broken chondrules; another fraction may be fragments of coarse-grained, high-temperature materials. The bulk consists of fine-grained, low-temperature nebular material that escaped the chondrule-forming process. Some authors restrict matrix to refer only to this fine-grained material.

The "classic" chondrule is spheroidal and shows evidence for formation as a molten droplet, thus implying formation temperatures at or above its melting temperature. In fact, only a small fraction of meteoritic chondrules have this classic shape. Because most chondrules have the mafic minerals olivine or pyroxene as the dominant mineral, minimum melting temperatures of 1600 to 1800 K are inferred, the values increasing with increasing content of olivine and with decreasing $FeO/(FeO + MgO)$ ratio in the mafic mineral. Recent studies show that some chondrules were incompletely molten, indicating maximum temperatures in the range of 1400 to 1600 K. Although some authors include as chondrules the igneously formed refractory inclusions observed in Allende and other CV chondrites, these inclusions formed substantially earlier than the true chondrules did, and they may have been melted by a different mechanism. For this reason, it simplifies discussions not to designate them as chondrules.

Chondrules come in a variety of compositions and textural types. The same types of chondrules are found in the different chondrite groups, but the relative abundances vary. Compositions of chondrules also vary from clan to clan. The differences are relatively small, and it is probable that the same basic chondrule-formation process(es) occurred throughout the nebula. Because ordinary-chondrite chondrules have been studied much more often than those of other groups, it is convenient to focus our discussions on the studies of this set.

Figure VII-3 shows nine chondrules and illustrates the variety of textures and common mineralogical compositions encountered. All except the chondrule in Figure VII-3a are from ordinary chondrites. Those in Figure VII-3c,g have the "classic" circular outline; in fact, however, the irregular shapes of the remaining chondrules are more typical. Six of the chondrules are porphyritic (a **porphyritic** rock consists of well-defined, coarse mineral grains set in a fine-grained or glassy host material often designated mesostasis), in keeping with the observation that porphyritic chondrules are more abundant than the barred olivine or radiating pyroxene chondrules most-commonly illustrated in texts on meteorites.

Metal is relatively uncommon in chondrules; in this set, it is mainly recognizable as spherules in Figure VII-3i. It is also present in the chondrule of Figure VI-3h, in which it has precipitated as clouds of micrometer-sized grains inside angular olivine grains. This latter metal is nearly free of nickel, indicating formation by reduction of FeO (in the olivine) to metallic iron. Textural evidence indicates that this reduction cannot have occurred after formation of the chondrule, so these olivine grains containing the dusty metal must be relicts that were not melted during the formation of the chondrule.

Surveys of unequilibrated ordinary chondrites yield the following relative abundances of their various millimeter-sized constituents: barred olivine chondrules, 3 to 4 percent; radiating pyroxene chondrules, 7 to 9 percent; porphyritic chondrules, 75 to 85 percent; glassy or slightly devitrified chondrules, 5 to 10 percent. Many chondrules have textures intermediate between these general types, so these abundances must be taken *cum grano salis.*

There is overlap in composition as well as in texture among the different chondrule classes. This observation suggests a single basic formational mechanism; modest variations in certain conditions can account for the observed ranges. For example, if a molten droplet is cooled rapidly, it may quench to a glass; the same droplet cooled at an intermediate rate may nucleate and grow crystals of the most-refractory minerals with the less-re-

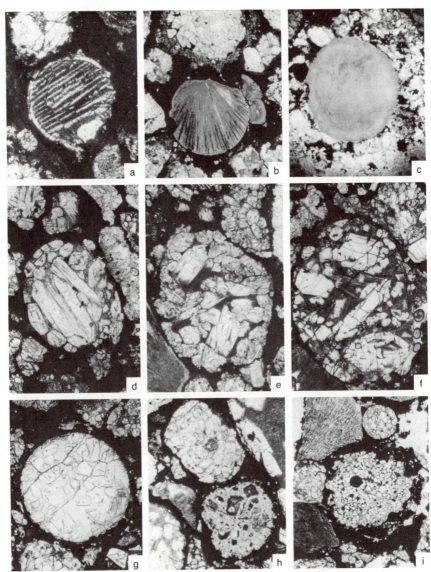

Figure VII-3. Textures of typical chondrules. Numbers of UCLA and Smithsonian Institution (SI) thin sections are listed for documentation purposes. *(a)* Barred olivine chondrule. Parallel plates of olivine are surrounded by a nearly continuous, spherical olivine shell. The interstitial fine-grained material is probably devitrified glass (Allende (CV3), SI 4744-1). *(b)* Radial pyroxene chondrule. Thin crystals of pyroxene radiate from a point near the chondrule's surface, giving the appearance of a Japanese fan. The opaque phase between these crystals is troilite. One large and several small indentations are visible on

fractory material remaining glassy; an initially molten droplet cooled very slowly will crystallize in its entirety. Additional textures can result from these three types of cooling histories if the initial mass was not entirely molten.

The temperatures required to melt chondrules are so high that all elements are volatile; during the moment of melting, evaporation from the surface must have occurred. Evaporation and radiation quickly cool the surface so that the common elements are no longer volatile, and volatilization loss depletes a thin surface layer of the substances that remain volatile at this lower temperature. If the cooling rate is slow enough, additional volatiles will diffuse from the interior to the surface and evaporate there. Somewhat less likely is that the interior might convect and thus very efficiently transport volatiles to the surface. In this latter case, one would expect to see a systematic depletion of volatiles in chondrules. No such systematic effect is present, though a few ($\sim$ 10 percent) of the chondrules do show volatile depletions by as much as a factor of 2. Some others are even slightly (up to 30 percent) enriched in volatiles. Thus it appears that most chondrules cooled very rapidly.

The only elemental classes that are systematically low in chondrules are the siderophiles and, to a lesser degree, **chalcophiles**, elements geochemically associated with FeS. The missing metal and sulfide is mainly present in

---

the chondrule's surface. Such "craters," common in chondrules of this type, may be formed either by low-velocity collisions or by post-solidification loss of metal/sulfide droplets from the surface (Semarkona (LL3), SI 1805-1). *(c)* Cryptocrystalline chondrule. This chondrule is composed of an intergrowth of extremely fine-grained pyroxene crystals and minor amounts of mesostasis (Enshi (H4), UCLA 324). *(d)* Porphyritic pyroxene chondrule. Large phenocrysts of pyroxene are surrounded by smaller crystals, all enclosed by a finely crystalline mesostasis (Semarkona (LL3), SI 1805-4). *(e)* Porphyritic olivine–pyroxene chondrule. Large phenocrysts of pyroxene enclose small, anhedral grains of olivine in a dark, very fine-grained mesostasis (Manych (LL3), UCLA 295). *(f)* Porphyritic olivine chondrule. Large olivine phenocrysts are surrounded by a mesostasis consisting largely of fine pyroxene needles. The olivine crystals have deep embayments and may have been partially resorbed during the formation of the surrounding melt (Semarkona (LL3), SI 1805-4). *(g)* Porphyritic olivine chondrule. Euhedral olivine phenocrysts are surrounded by abundant, colorless glass (Semarkona (LL3), SI 1805-1). *(h)* Two porphyritic olivine chondrules. Both of these chondrules contain dark as well as clear olivine phenocrysts. The dark appearance of these crystals is due to the presence of extremely fine grains of low-nickel metal (Semarkona (LL3), SI 1805-4). *(i)* Porphyritic olivine chondrule. Small phenocrysts of olivine and droplets of metal are surrounded by transparent brown glass (Semarkona (LL3), SI 1805-1). (Photos by J. N. Grossman and A. E. Rubin.)

the matrix as small grains that are generally not spheroidal enough to be designated chondrules. The uniform "solar" composition summarized in Figure VII-4 is one important property that must be explained by a successful model of chondrule origin.

Another key property of chondrules is their very limited size range at each formation location. Chondrules in the enstatite, ordinary, and refractory-rich clans are rarely outside the range of 0.2 to 2 mm, and the analogous range in the minichondrule clan is about 0.05 to 0.5 mm. The two

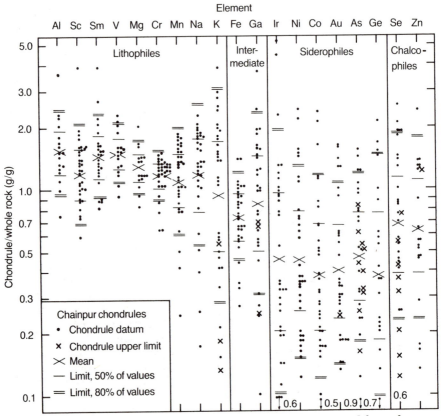

Figure VII-4. Chondrite/whole-rock ratios in chondrites separated from the Chainpur chondrites; the two indicated ranges for each value include 50 and 80 percent, respectively, of the values for each element. Within each geochemical category, elements are arranged in order of increasing volatility to the right. Lithophile ratios are uniformly high and siderophile ratios uniformly low, both reflecting an underabundance of metal in chondrites relative to the whole rock. In each category, ratios appear to decrease slightly with increasing volatility. (Diagram by J. Grossman.)

extreme possibilities are (1) that the observed size range accurately depicts that produced during the chondrule-formation process, or (2) that the formational process produced a wider range that has been reduced to the observed distribution as a result of selectional biases. Aerodynamic selectional effects have been proposed.

As discussed in Chapter II, in materials in which metallic Fe–Ni coexists with silicates, the $FeO/(FeO + MgO)$ ratios of the mafic minerals olivine and pyroxene are measures of the degree of oxidation of the system. In the most unequilibrated chondrites, chondrule $FeO/(FeO + MgO)$ ratios often range from 1 to 40 mol%, and higher values are sometimes found.

Each of the most actively discussed models of chondrule origin can be assigned to one of the following categories:

1. condensation in a monotonically cooling, unfractionated nebula;
2. condensation in a fractionated nebula subjected to a sharp temperature perturbation;
3. condensation in the atmosphere of a (hypothetical) protoplanet (a miniature solar nebula);
4. melting of solid grains by transient events (such as lightning) in the nebula;
5. melting of solids by impacts on the surface of a large parent body;
6. melting of solids by impacts between very small bodies in the nebula;
7. melting of interstellar materials accreting into the solar nebula.

Unfortunately it is easier to find fault with these models than it is to praise them.

Liquid silicates are unstable at all temperatures in a low-mass nebula (one in which H_2 pressures are 10^{-3} atm or less). Equilibrium condensation of liquids in a monotonically cooling nebula is possible only if p_{H_2} is much higher (at least 1 atm), as expected in a nebula having an implausibly high extrasolar mass greater than $1 M_\odot$, or if the nucleation of liquid droplets occurs earlier (at higher temperatures) than the nucleation of solid grains — that is, if kinetic factors dominate over thermodynamics. Even if either of these conditions prevailed, equilibrium with nebular gases at the high temperatures necessary to prevent liquids from crystallizing should have led to highly reduced and volatile-free chondrules, containing negligible amounts of FeO and of volatiles (including sulfur). Such gross depletions of volatiles are not observed in Figure VII-4. The possibility has been raised that the observed volatiles constitute a late exterior veneer, but experiments in which the surface has been removed showed that large fractions (more than 50 percent) of the volatiles are in chondrule interiors.

The high mafic $FeO/(FeO + MgO)$ ratios often found in unequilibrated chondrules (see Figure VII-5) are especially difficult to explain with a condensation model. At the high nebular temperatures required to produce a

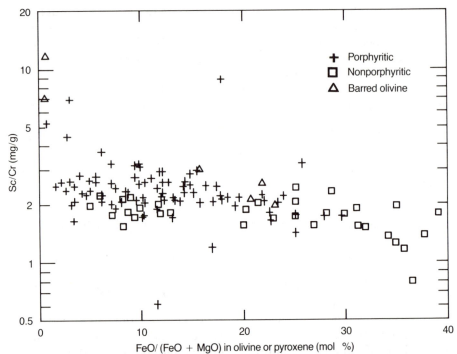

Figure VII-5. The wide range of **FeO/(FeO + MgO)** contents in olivine and pyroxene are inconsistent with condensation models of chondrule origin. The ratio of a refractory element (scandium) to a common element (chromium) decreases with increasing FeO/(FeO + MgO) ratio, as expected if chondrules formed by the mixing of two main lithophile components: a high-temperature component having high refractory and low FeO contents, and a low-temperature component having low refractory and high FeO content. (Diagram from J. Grossman and J. Wasson, in *Chondrules and their Origins*, ed. E. King, p. 88, Lunar Planet. Inst., Houston, 1983.)

liquid, the iron should be present as Fe – Ni having the solar nickel content of ~5.5 percent (no taenite should be present). Instead, some metal grains in chondrules are observed to have much-higher nickel concentrations (see Figure VII-6), inconsistent with condensation.

In fact, it is not high H_2 pressures that are needed to stabilize liquid silicates and metals, but instead high pressures of the gaseous phases containing magnesium, silicon, and iron. One proposed mechanism for achieving this enrichment in a low mass nebula is the following: (1) solid grains condense and migrate to the nebula midplane (see Chapter VIII); (2) a thermal event of unspecified origin evaporates the grains, leading to a marked enhancement of the partial pressures of the gaseous species of

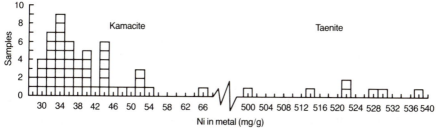

Figure VII-6. The nickel concentrations in the metal grains inside chondrules in the highly unequilibrated Bishunpur LL chondrite are distinctly different from those in the metal in equilibrated ordinary chondrites. Mean nickel concentrations in kamacite are ~40 mg/g, much lower than the 60 to 65 mg/g in equilibrated ordinary chondrites; central nickel concentrations in taenite are uniformly high (~500 mg/g or greater) compared to typical values of 300 to 400 mg/g in equilibrated chondrites, and the nickel is unzoned. The metal in equilibrated chondrites seems to have formed by slow cooling from high temperatures (~950 K or higher); the formation of the low-Ni kamacite in unequilibrated chondrites may in part have involved the reduction of iron from FeO during chondrule formation; the taenite may have formed by poorly understood processes at low temperatures. (Data from E. Rambaldi.)

magnesium, silicon, and iron (and more-refractory elements); (3) recondensation in the midplane produces chondrules as liquid droplets. This model shares the problems already noted for condensation models; it does not account for high FeO/(FeO + MgO) ratios in the silicates or for high nickel content in the metal. Furthermore, although it avoids the necessity for metastable (kinetically constrained) condensation of liquids, it requires a transient heat source many times more intense than that needed to melt preexisting grains without vaporizing them.

If a high-mass nebula did form, it might lead to the formation of gaseous protoplanets (resembling Jupiter but ~10 times smaller) in the inner solar system. It has been suggested that chondrules could form as liquid droplets in the atmosphere of a cooling protoplanet. Such chondrules, however, should be highly reduced and have low volatile contents, as expected also from the two preceding condensation models. Further, silicate condensates should immediately settle to the solid or liquid "core" of the planet, and there appears to be little likelihood that the eventual dispersal of the gaseous shell would lead to an appreciable introduction of chondrules into the nebula.

Turbulence in the nebula can produce a buildup of static charge culminating in lightning, particularly as a result of shear between the dusty midplane and the gas-rich regions above and below. If the lightning melted random collections of grains including all intermediate categories between

(1) coarse (millimeter-sized), low-$FeO/(FeO + MgO)$ grains that stopped equilibrating with nebular gases at high temperatures and (2) fine, high-$FeO/(FeO + MgO)$ grains that equilibrated down to low temperatures, then the observed ranges in mafic compositions and volatile contents could be explained.

Figure VII-5 shows a trend that supports such a mixing model; in the chondrules from unequilibrated ordinary chondrites, contents of refractory elements such as scandium and samarium (here normalized to chromium to allow for dilution by nonlithophile components) decrease with increasing $FeO/(FeO + MgO)$ ratios in the mafic minerals. This effect is not predicted to occur during nebular condensation, but seems to require the mixing of a refractory-rich, high-temperature component having a low $FeO/(FeO + MgO)$ ratio with a refractory-poor, low-temperature component having a low $FeO/(FeO + MgO)$ ratio.

If the conversion of turbulent energy to lightning and the application of the energy in the lightning to the heating and melting of dust was relatively inefficient, then the fraction of remelted grains would be smaller than the observed fraction of chondrules. However, it appears that an efficiency of about 10^{-6} for the conversion of turbulent energy to chondrule heat energy is all that is needed to melt the solids within a $\sim 10^{5}$-yr period, and this value does not appear to be implausibly high.

Several authors have endorsed the idea that chondrules could be produced by impacts into the regolith of a relatively large, atmosphereless body. Such a model shares the advantage of the other remelting models that the variety of chondrule compositions can be produced by making the plausible assumption that the initial material equilibrated with nebular gases over a wide range of temperatures. However, the samples we have of the regolith of one atmosphereless body, the Moon, contain very few spheroidal "chondrules"—no more than a few percent of the regolith, even if allowance is made for fragments that when sectioned show a circular outline on a portion of their surface. It has been suggested that the proportion of chondrulelike objects would be higher if the starting material were more porous than that on the Moon because, the greater the compressibility, the greater the amount of heat deposition during an impact shock. However, the porosity of the lunar regolith is high (~ 33 percent), and the porosity of a hypothetical parent body could not remain appreciably higher than this value during an extensive period of bombardment. The model also appears to predict that objects originating on small bodies or in the interiors of large bodies would contain no chondrules but would otherwise be identical in composition to the chondrule-bearing chondrites, whereas such chondrule-free ordinary chondrites have never fallen. A final problem concerns the relative impact velocities of ~ 3 km $\cdot$ s^{-1} required to melt the projectile and a minor fraction of the target. Such high interobject velocities are not predicted for nebular materials until *after* the planets have formed and, because they exceed asteroidal escape velocities (see Equation IV-1), one would expect net destruction of the asteroid rather than growth.

One reason there are so few chondrulelike bodies on the Moon is that the impact melt tends to get mixed with fragmental regolith dust to produce so-called **agglutinates**. It thus appears that the fraction of chondrules would be much higher if the impacts were to occur between small (less than 10 cm) objects in space. Because chondrules are produced before and during the agglomeration of grains to form rocks, the fraction of chondrules should be essentially independent of the size of the final object. On the other hand, as discussed more fully in Chapter VIII, the minimum impact velocity of 3 km · s^{-1} is much greater than that expected at the end of the period of condensation. In order to agglomerate rocks by means of gravitational collapse, the mean relative velocities must be less than 0.1 km · s^{-1}. This conflict poses a great difficulty for this model.

The formation of the solar nebula did not occur instantaneously; accretion of interstellar material to the nebula occurred as long as the solar system remained in its placental cloud. After the solar nebula had grown to a sizable fraction (say, 10 percent) of its final size, interstellar grains accreting to it would enter the nebula with velocities exceeding 10 km · s^{-1}, and they would be heated by friction with the nebular gases. The phenomenon would be completely analogous to the heating of meteoroids as they pass through the upper atmosphere of the Earth.

Those solids that formed in interstellar clouds probably condensed at low temperatures from a roughly solar mixture of the elements, and thus they should have most iron present as oxides or sulfides, but some of the interstellar solids may be highly reduced material that condensed at high temperatures in the atmospheres of stars. There is no way to estimate the relative fractions of these two classes of materials, but it seems likely that the majority was oxidized materials of interstellar origin.

If the interstellar matter fell into the nebula at times and/or places of low temperatures, the liquid would freeze as soon as it reached terminal velocity, and the mafic-mineral FeO/(FeO + MgO) ratio in the resulting chondrule would probably reflect the ratio in the interstellar precursor. In contrast, if nebula temperatures were high, then the liquid might survive for a longer period before crystallizing, and low mafic FeO/(FeO + MgO) ratios would result from an approach to equilibrium between the liquid droplet and the nebular gases. Because even formation in a low-temperature nebula will result in the material being in the liquid state for several seconds, some volatile loss should occur in all cases.

In order to account for the observation that as much as 90 percent of the lithophiles in some unequilibrated ordinary chondrites are present in chondrules, the chondrule-formation process must have been very efficient in some locations; most of the matter that migrated to the nebula midplane and agglomerated there must have consisted of melted interstellar grains. This conflicts with the arguments presented earlier in this chapter that interstellar matter was largely vaporized in the inner solar system. A possible solution to this dilemma would be to assume that most matter vaporized but that most of the vaporized material then recondensed as fine-grained materials that remained suspended in the nebular gas.

It is commonly held that interstellar materials are fine-grained, low-temperature "smoke" particles. If all were of this composition, an interstellar chondrule model has difficulty in accounting for the range of chondrule compositions shown in Figures VII-4 and VII-5.

In summary, there is no completely satisfactory model of chondrule origin. None of the condensation models can account for the high $FeO/(FeO + MgO)$ ratios commonly found in mafic minerals; neither alternative (high nebular pressures or kinetic constraints on the nucleation of solids) appears to provide a satisfactory solution to the problem. Among the impact models, the regolithic version has many problems; the small-bodies-in-space model has fewer, but one of its problems (the need for impact velocities greather than $3 \text{ km} \cdot \text{s}^{-1}$) appears to be fatal. During the past decade, the lightning model was all but buried by meteorite researchers, but there is some hope that additional theoretical work can lead to resuscitation. The model calling for the accretionary melting of interstellar grains may be viable, but there are numerous potential problems, and a detailed model has yet to be generated. This is a field that still requires much additional research.

Suggested Reading

Anders, E. 1971. Meteorites and the early solar system. *Ann. Rev. Astron. Astrophys.* 9:1. A technical review of meteoritic evidence regarding the solar nebula with emphasis on fractionation processes.

Dodd, R. T. 1981. *Meteorites: A Petrologic-Chemical Synthesis.* University of Cambridge Press. Pages 61–68 and 121–131 review chondrule origin at a moderately technical level.

Grossman, L., and J. W. Larimer. 1974. Early chemical history of the solar system. *Rev. Geophys. Space Phys.* 12:71. A technical review of nebular condensation processes and their relationship to elemental fractionations in meteorites.

Wasson, J. T. 1978. Maximum temperatures during the formation of solar nebula. In *Protostars and Planets*, ed. T. Gehrels, pp. 48–501. University of Arizona Press. A discussion of the meteoritic evidence regarding maximum nebular temperatures.

Wood, J. A. 1979. *The Solar System.* Prentice Hall. A description of nebular processes is presented in Chapter 7.

CHAPTER VIII

Differences Between and Within Chondrite Groups: Evidence Regarding Nebula-Fractionation Processes

As discussed in Chapter II, the chondrite groups differ systematically in their bulk compositions. For example, the abundance of refractory elements are highest in the CV chondrites and lowest in the EH and EL chondrites. Siderophile elements (such as nickel, gold, and iridium) also are fractionated, with the highest abundances in the CI, H, and EH chondrites and the lowest in the LL and EL chondrites.

As noted in Chapters II and VII, evidence of various sorts shows that the chondrites formed in the solar nebula. Although many experienced a later reheating to moderate temperatures (1000 to 1200 K), there was no loss of major phases during this reheating, and thus there was no way in which refractories or nonvolatile siderophiles could have been fractionated. Therefore, the processes that fractionated the refractories and siderophiles must have occurred in the solar nebula.

It is fascinating to try to put together a plausible picture of how these processes were generated in the nebula. To do this, cosmochemists must carefully consider the key features of astrophysical models describing the processes that led to the formation of the Sun.

Even more interesting is to turn the collaboration with the astrophysicists around — to test suggested solar nebular models by their ability to account for the observed fractionations. For example, some current models call for large amounts of turbulent mass transport through the nebula, whereas others call for little turbulence and little transport. Some models call for low (less than 1000 K) maximum nebula temperatures, some for temperatures hot enough to vaporize silicates. Careful examination of the properties recorded in the chondrites should show which classes of models are more and which are less plausible.

Processes in the Protoplanetary Solar Nebula

As discussed at the end of Chapter VI, the solar system probably formed together with many other stars in a dense interstellar cloud similar to that observed in Orion. The compressive trigger that initiated star formation was probably either a supernova shock wave or a standing gravity wave associated with a galactic arm.

After a particular region had contracted in linear dimensions by about an order of magnitude, subregions became gravitationally unstable, and fragmentation occurred. This process of fragmentation continued as the cloud contracted. We cannot determine how many stars formed in the Sun's region; a reasonable estimate might be 10 to 1000. The total linear contraction of Sun-sized parcels during this stage was by a factor of $\sim 10^4$.

The advent of computers with very large memories and very short processing times has allowed the development of complex and detailed models of star formation. These models offer numerous insights into the kinds of interstellar and nebular processes that may have affected primitive solar-system matter. Nonetheless, uncertainties regarding the conditions in interstellar clouds prevent the accurate definition of initial conditions (particularly the distribution of angular momentum) that played a major role in determining the properties of the solar system. Furthermore, it has not proven possible to model the turbulence in the cloud or nebula, to allow for episodic accretion of additional material to the nebula, and (in most cases) to follow the calculations through until the central region forms a main-sequence, hydrogen-burning star.

Because these uncertainties lead to poorly constrained conditions in the solar nebula, detailed numerical models of the various processes occurring during the evolution of the nebula have not been attempted. Numerical calculations have been made for individual accumulation processes, and these results have been incorporated into a qualitative sketch of nebula evolution.

Table VIII-1 lists the processes that lead from an interstellar cloud to the formation of planets and the meteorite parent bodies. This list includes only those processes that are currently accepted as having played important roles; most processes assigned minor importance are not included. The lower extremes on the time ranges for steps 2, 3, and 4 are probably unrealistic; they hold only for the case that the accretion of the nebula is completed within a time substantially less than 10^4 yr.

As discussed at the end of Chapter VI, a great deal of heat was liberated by the collapse of an interstellar cloud to form the solar system. In Chapter VII, I discussed chondritic evidence indicating the maximum temperatures reached by the nebula; these temperatures were at least 1400 and 1460 K at pressures of 10^{-6} and 10^{-5} atm, respectively.

The temperature at each nebular location depended mainly on the rate of input of gravitational energy and the temperature-dependent rate of loss by radiation. Most star-formation models yield temperatures much lower than those inferred in Chapter VII, and some researchers are therefore searching for ways to produce high-temperature assemblages in a low-tempera-

Table VIII-1
Processes occurring between the contraction of an interstellar cloud and the solar-nebula formation of compact solid bodies and finally planets

	Process	Time required
1	Collapse of a cloud from $\sim 10^5$ AU to ~ 10 AU; liberation of gravitational energy leads to partial evaporation of interstellar solids.	$\sim 10^6$ yr
2	As collapse ends, the nebula cools, and evaporated matter starts to recondense.	$\sim 10^4 - 10^6$ yr
3	Cooling continues; the interiors of large grains cease to equilibrate with the gas.	$\sim 10^4 - 10^6$ yr
4a	Some solid matter settles to the median plane; partly because of inelastic collisions, the particles in some regions undergo gravitational contraction to form meter- to kilometer-sized planetesimals.	$\sim 10^4 - 10^6$ yr
4b	Some fine matter remains suspended in the gas and does not form planetesimals.	
5	Planetesimals collide with one another, leading to coagulation to larger bodies and compaction to form bodies as dense and strong as chondritic meteorites.	$\sim 10^5$ yr
6	As a result of mutual perturbations, orbits become more eccentric, and bodies sweep up larger and larger volumes until planets are formed.	$\sim 10^8$ yr

ture nebula. However, it appears at least equally satisfactory to modify the star-formation models to yield higher temperatures — for example, by reducing the initial angular-momentum density assumed in the calculations. It seems plausible that the regions in the cloud that undergo collapse are precisely those having anomalously low angular-momentum contents. Temperatures also would have been enhanced if there were a dusty region immediately around the solar nebula (perhaps at a distance of 10^4 AU from the Sun). The evidence that the maximum nebula temperatures really did reach at least 1400 K at ~ 3 AU from the Sun presented in Chapter VII seems strong, and I will use this as my working assumption.

The maximum duration of the high-temperature period in nebular history is also known. Because ^{26}Al was preserved in some samples, the period cannot have exceeded 10^6 yr. A minimum estimate is set by the time for cold matter of uniform density to collapse. The following formula is based on a closed-system collapse (one without mass loss or gain) of matter having uniform density and a spherical shape:

$$t = \frac{\pi}{2} \left(\frac{R^3}{2GM} \right)^{1/2} \qquad \text{(VIII-1)}$$

where t is the free-fall time of the outermost matter, R is the radius of the cloud fragment in centimeters, M is the mass of the system in grams, and G (the gravitational constant) is $6.67 \times 10^{-8} \, cm^3 \cdot g^{-1} \cdot s^{-2}$. The initial radius of the cloud fragment that formed the solar system was about 4×10^4 AU, and we are interested in the time necessary for this fragment to reach a mean radius of ~ 3 AU. The starting radius corresponds to a mean hydrogen-atom number density of $10^3 \, cm^{-3}$ and a mass of $1.1 M_\odot$. These parameters yield a free-fall time of 1.3 Myr. If we hold the mass at $1.1 M_\odot$ but start the collapse at $R = 4 \times 10^3$ AU, corresponding to a hydrogen-atom density of $10^6 \, cm^{-3}$, then the free-fall time is reduced to 41 kyr.

This model is oversimplified for the following reasons.

1. The matter possessed angular momentum, which slowed the collapse.
2. The matter was probably not uniformly distributed; it seems likely that the central region started with a higher density, perhaps reflecting density fluctuations in the cloud or even the presence of a sizable ($0.0 M_\odot$ to $0.5 M_\odot$) body that had formed in but escaped from a neighboring stellar system.
3. Random motions in the cloud probably led to the infall of some matter at rates faster than the free-fall rate, and they also led to the addition of matter to the solar nebula over an extended period. The typical period during which a star-forming region might survive appears to be a few megayears, the lifetime of an O-type star, because the high radiation intensities from these stars lead to cloud dissipation. A suggested lower limit on the duration of the high-temperature period is about 10 kyr.

I noted earlier that the temperature at each location depends on the balance between infall rate and the rate of outward leakage of radiation. Because these rates will have varied with time, the temperature must also have varied. The rate of temperature change affects properties such as grain size or the diffusion of volatile elements into a refractory host grain, and it seems possible that the nebular cooling rate could be inferred from a study of the sizes and volatile distributions of matrix grains in the most unequilibrated chondrites.

The gaseous solar nebula had an angular momentum higher than that of the present solar system because the angular momentum per gram of the material that was later expelled was almost certainly greater than that in the remaining material. Initially, the protoplanetary materials moved in heliocentric orbits having inclinations that were nearly randomly distributed. However, there was a net angular momentum in the nebula and, as a result, the nebula eventually took the form of a relatively thin disk consisting of material moving about the Sun in the same direction. Most of the matter in the disk was gaseous, and most of the gas was hydrogen (~ 781 mg/g) or helium (~ 199 mg/g); all other elements account for only 20 mg/g (see Table VIII-2).

The variation in pressure and in the related quantity **surface density** can be estimated from the masses and compositions of the planets. (The **surface density** is the mass per unit area in a section perpendicular to the nebular plane.) The composition of each planet is estimated (see the more extensive discussion in Chapter IX), and the minimum amount of material in that part of the nebula is then calculated from the amount of the element having the highest planet/Sun abundance ratio. In the terrestrial planets, that element is iron; in the giant planets, it is probably oxygen, silicon, magnesium, or iron but, because the amounts of these are highly uncertain, a still lower "minimum" for Jupiter and Saturn can be estimated from the assumption that their compositions are solar. The amount of other elements is calculated on the basis of solar abundances, and the total matter is distributed

Table VIII-2
Atomic abundances of the 14 most-abundant elements in solar matter, and approximate partial pressures of the 11 most-abundant gases in a solar nebula having $p_{H_2} = 1.0 \times 10^{-4}$ atm and $T = 1500$ K

Element	Atomic abundance	Gas	Pressure (atm)
H	2.2×10^{10}	H_2	$\equiv 1.0 \times 10^{-4}$
He	1.4×10^9	He	1.3×10^{-5}
O	1.5×10^7	CO	8.5×10^{-8}
C	9.3×10^6	H_2O	4.5×10^{-8}
Ne	3.4×10^6	Ne	3.1×10^{-8}
N	2.0×10^6	NH_3	1.8×10^{-8}
Mg	1.1×10^6	Mg	1.0×10^{-8}
Si	$\equiv 1.0 \times 10^6$	Fe	7.9×10^{-9}
Fe	8.7×10^5	SiO	6.4×10^{-9}
S	4.9×10^5	SiS	2.7×10^{-9}
Al	8.6×10^4	H_2S	5.0×10^{-10}
Ca	6.1×10^4		
Na	5.6×10^4		
Ni	4.9×10^4		

into a band that extends from a heliocentric distance intermediate between the semimajor axes of the planet and its inner nearest neighbor outward to a distance intermediate beween the semimajor axes of the planet and its outer nearest neighbor.

Figure VIII-1 shows the variation in surface density σ calculated in this fashion. The limits on the band of each planet are taken to be the geometric mean of the semimajor axes of adjacent planets. (To calculate a geometric mean of two numbers, one multiplies the numbers and takes the square root of the product. Thus the inner edge of the Earth's zone is 0.84 AU, obtained by taking the square root of the product of Earth's semimajor axis 1.00 and Venus' semimajor axis 0.70.) Two curves are shown for surface density, based on different estimates of the nebular mass in the Jupiter and Saturn; the lower estimate (σ_A) is based on the estimated amounts of hydrogen, and the higher (σ_B) is based on estimates that the masses of their rocky cores are 12 to 25 times higher than expected on the basis of hydrogen contents. Each curve passes through the values estimated from the masses of Venus and Earth, based on assumed bulk iron contents of 300 mg/g for each planet. Note that the surface density estimated from the mass of Mercury is ~5 times lower than that predicted by the curves; that for Mars is lower by a factor of at least 20. This discrepancy is generally interpreted to indicate inefficient accretion of these planets as a result of gravitational stirring by the Sun (in the case of Mercury) or by Jupiter (in the case of Mars). The mass loss from the asteroidal belt has been much greater yet.

Central nebular pressures (p_A and p_B) as a function of the heliocentric distance are also shown in Figure VIII-1. These values are based on the surface density and the thickness of the nebula. The density of the nebula decreases with height above the median plane. It is convenient to define a disk of thickness D that contains one-half of the mass. This thickness depends on temperature T and distance a from the Sun[1]:

$$D = 0.08 \left(\frac{T}{1000} \right)^{1/2} a^{1/2} \qquad \text{(VIII-2)}$$

where T is expressed in K, and D and a in AU. The central H_2 pressure p_{H_2} is approximately

$$p_{H_2} = 5 \times 10^{-5} \frac{\sigma}{D} \qquad \text{(VIII-3)}$$

---

[1] Goldreich and W. R. Ward, *Astrophys. J.* 183:105, 1973.

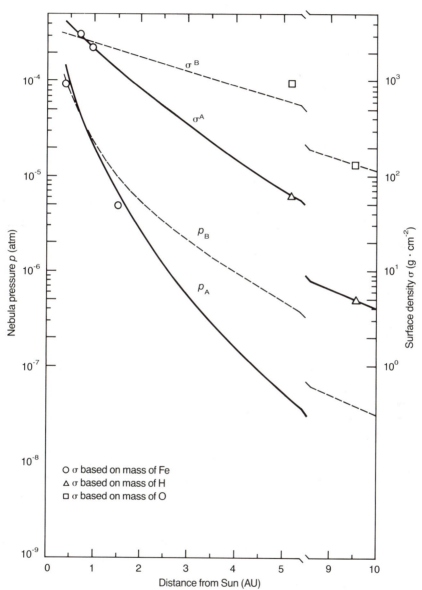

Figure VIII-1. Minimum surface densities and pressures in the solar nebula, estimated from the masses and compositions of the planets. Two estimates are used for Jupiter and Saturn. See text for details. Large amounts of nebular solids have been lost from the formation regions of Mercury, Mars, and the asteroids.

where p_{H_2} is expressed in atm, and the surface density σ in g $\cdot$ cm^{-2}. The pressure at the top and bottom of this region of thickness D is ~ 0.7 times that at the center; 90 percent of the nebular gas is in this layer. Thus it is not unreasonable to assume in condensation calculations that the nebular pressure is constant.

Near the Sun, the condensable matter was largely gaseous at the time maximum temperatures were reached. The elemental composition of this gas was almost certainly solar, identical to that in the solar atmosphere today except for abundances of a few light elements that have been altered in the outer regions of the Sun by nuclear reactions. Table VIII-2 lists abundances of the 14 most-abundant elements (excerpted from Appendix D) and the partial pressures of the 11 most abundant gaseous species in a nebula having an H_2 pressure of 1.0×10^{-4} atm and a temperature of 1500 K. The gas pressures are calculated by techniques discussed in Appendix F. As discussed in Chapter II, it is difficult to determine most elements in the solar atmosphere to a precision better than ± 30 percent, and Table VIII-2 is therefore based on CI chondrite abundances (also listed in Appendix D) for those elements that completely condensed.

As the nebula cooled, the elements condensed, starting with the most-refractory elements. If sufficient thermodynamic data are available, it is a straightforward task to calculate the temperature at which each element condensed under equilibrium conditions; (see Appendix F). Appendix G lists condensation temperatures above ~ 400 K in nebular regions having total $(=H_2)$ pressures of 10^{-4} to 10^{-6} atm. Most of the listed 50%-condensation temperatures should be correct to within $\pm \sim 20$ K, but a few may have much-larger errors because key data were not available.

During condensation, the common elements form their own minerals, whereas the trace elements are distributed among these phases. With decreasing temperature, the solubilities of the trace constituents in the main host phases generally decrease, and the diffusion coefficients that control the transport of trace constituents into the interiors of the host minerals decrease exponentially. As a result, the more volatile the trace element, the more likely it is to be localized on grain surfaces — or even to form its own (probably very-fine-grained) phase, or to condense together with a few other elements having similar volatilities as fine, amorphous "smoke" particles. After chondritic grains have condensed, the mass of gas is 300 times greater than that of the grains.

These grains experience a drag force that tends to bring them toward the median plane of the nebula and nearer to the Sun. The motion of large grains is dominated by the gravitational attraction of the Sun, and so the grains tend to follow "Keplerian" orbits about the Sun. In the absence of gas, a grain above the median plane would follow an inclined elliptical orbit that would carry it through the median plane twice during each orbit. Much of the motion of gas molecules is controlled by elastic collisions among these molecules. A gas molecule above the median plane will, in the absence of turbulence, be held by collisions with other molecules at essentially the same position above the plane throughout a circuit about the Sun. Thus, the gas molecule tends to move in a plane parallel to the median plane of the nebula. Because there is an outward pressure gradient in the nebula, the gas molecule moves through its orbit at a velocity less than the Keplerian orbital velocity.

As a result of the friction, the grain exchanges angular momentum with the gas; because its velocity is reduced, the grain spirals toward the Sun. Its inclination is also reduced, and thus its orbit is shifted toward the median plane. When 10 percent of the particles have settled into a layer having a thickness $\sim 10^{-4}$ that of the gas, the mass of the grains is greater than that of the gas, and the gas and grains in this region rotate together at roughly Keplerian velocities. There is no further tendency for the semimajor axes of the settled grains to decrease.

It is interesting to compare the behaviors of grains having radii of 1 mm and 1 μm — sizes typical of chondrules and interchondrule matrix, respectively. At $p_{H_2} = 10^{-5}$ atm and $T = 1000$ K, the settling time t_{set} is approximately[2]

$$t_{set} \cong 10^3 r^{-1} \qquad \text{(VIII-4)}$$

where the grain radius r is expressed in mm, and t_{set} in yr. Thus, a 1-mm grain settles to the median plane in $\sim 10^3$ yr, whereas a 1-μm grain requires $\sim 10^6$ yr. During these periods, the 1-mm particle would migrate ~ 0.002 AU toward the Sun, whereas the 1-μm particle's motion is so strongly coupled to that of the gas that its sunward drift is negligible.

When the density of solids in the median plane has been increased to $\sim 4 \times 10^{-7}$ g $\cdot$ cm^{-3} through settling by a factor of 10^4 at $p_{H_2} = 10^{-4}$ atm or 10^5 at $p_{H_2} = 10^{-5}$ atm, the dusty layer fragments into regions that are self-gravitating. Although it is not yet clear which of several possible mechanisms is the dominant one, the particles in these regions should gradually dissipate their random interparticle velocities and contract to form solid bodies with masses of $\sim 10^{16}$ to 10^{18} g and densities of ~ 1 g $\cdot$ cm^{-3}, corresponding to radii of 1.4 to 7 km. This planetesimal formation is believed to occur within a period of 10^4 yr. Because this period is shorter than the settling time of 10^6 yr estimated for grains having radii of 1 μm, the first generation of planetesimals should be depleted in elements that have higher concentrations in the finer grain sizes, and rich in elements that have

---

[2] S. J. Weidenschilling, *Icarus* 44:172, 1980.

higher concentrations in coarser grains. We will return to this point in our discussions of the elemental fractionations observed among the various groups of chondrites. It is possible that an appreciable fraction of nebular solids never settled to the median plane. Some of these solids may have been swept up by asteroid- or planet-sized bodies, but most were probably evaporated and lost together with the nebular gases when these were dissipated, perhaps by an intense flux of photons or by a T-Tauri wind (see Chapters IV and VI). In this case, the surface densities estimated from the masses and compositions of the planets (Figure VIII-1) are lower limits. Some loss of solids must have occurred, and it is probably better to increase these surface densities by a factor of 2 when constructing minimum-mass nebular models.

This collapse of unstable dust-rich regions in the median plane appears to solve a problem that had vexed cosmochemists for several decades, that of finding a "glue" to hold particles together when they are much too small to have appreciable gravitational fields but are too large to be held together by static electrical charges. There are still some unanswered questions, however, regarding the perturbing effects of turbulence on this process, particularly in the inner solar system.

The planetesimals produced by the dust-collapse mechanism must have had low densities and little strength. If centimeter- or meter-sized chunks of such material were to enter the Earth's atmosphere, they would never fall as meteorites but would be disrupted to dust at very high altitudes (at least 80 km). To produce tough meteoroids from low-density planetesimals requires compaction and at least a minor amount of sintering (melting or recrystallization of grain surfaces to make adjacent grains fit together tightly and, in some cases, to produce chemical bonds between adjacent grains). The weak gravitational field of a kilometer-sized planetesimal cannot generate the observed low porosities (10 to 15 vol% pore space) or the sintering.

The most-plausible process for producing compaction and sintering is impact-induced shock. The setting is not clear, but a plausible scenario is that this shock occurred during the growth of kilometer-sized bodies to asteroids having dimensions of ~ 100 km. It seems clear that this growth was a necessary intermediate step in the sequence of processes that led from grains to planets. The amount of deposited shock energy per unit mass probably varied widely, ranging from minor recrystallization of grain surfaces in unequilibrated chondrites to extensive metamorphic recrystallization in the higher petrographic types.

The texture of the Leoville CV3 chondrite shown in Figure VIII-2 provides evidence for shock compaction. In this meteorite, the chondrules and refractory inclusions are preferentially oriented with their long axes parallel. A plausible scenario is that the sample was heated to a temperature of 1000 to 1200 K, at which the silicates became plastic; then a shear force

0   5   10 cm

Figure VIII-2. Section of the Leoville CV3 chondrite showing chondrules (gray) and refractory inclusions (whitish) that are ellipsoidal with their long axes parallel. The cracks in the section tend to be oriented perpendicular to the long axes of the chondrules and inclusions. The long diagonal of the section is 15 cm. (Photo by F. Wlotzka.)

was applied that elongated one axis of the previously spheroidal grains. This high-temperature period must have been brief, because Leoville minerals are highly unequilibrated; thus, the heating cannot have been produced by a long-lived heat source such as radioactivity. The only plausible mechanism for applying both heat and shear during a brief period appears to be to generate both by impact-produced shock.

Note that, if the porosities of solid materials produced by dust collapse are indeed high (much greater than 20 vol%), then none of the chondrites (including the volatile-rich CM and CI carbonaceous chondrites) are nebular products in the strict sense. All have undergone the compaction process; during this process, some grains were fragmented, and some fraction of the materials was subjected to elevated temperatures.

Because coarse grains settle to the median plane more rapidly than fine grains do, it follows that dust collapse could occur several times at a single location. If, as expected, different elements show different grain-size distributions, then the earliest-agglomerated materials were enriched in those elements more abundant in the coarser-grained fraction, whereas later generations were enriched in those elements more abundant in finer-grained materials. It is conceivable that some of these materials would be highly fractionated relative to solar abundances, and that therefore they might not immediately be recognized as nebular materials. Some suggestions have been made that certain highly fractionated meteoritic materials (such as the aubrites — see Chapter V) are of nebular origin, but there is not yet general agreement on this identification.

Fractionations Between and Among the Chondrite Groups

In Chapter II, several fractionations among the chondrite groups were discussed because the fractionated properties provided useful classificational information. Among these taxonomic parameters are refractory-element abundance, siderophile abundance, degree of oxidation, and oxygen-isotopic composition. We now return to these fractionations and also add volatile-element fractionations to the list.

As shown in Figure II-3, the abundances of refractory elements vary by a factor of 2.5, with the lowest values in EH and EL chondrites and the highest values in CV chondrites. Abundances of a larger set of refractory elements in the carbonaceous chondrite groups CI, CM, CO, and CV are shown in Figure VIII-3. These abundances are magnesium-normalized, but silicon normalization would result in trivial differences because the Si/Mg ratio is constant to within ~3 percent among the four groups. To facilitate comparisons, the data are shown as group/CI ratios. Lithophilic elements are plotted in the top portion of the diagram; siderophiles and others (especially "chalcophiles" that condense as sulfides) are plotted in the bottom portion. Within each set, the nebular condensation temperatures of the elements decrease to the right. Within each chondrite group, the eight refractory lithophiles (Al, Sc, Ca, Lu, Yb, Eu, Sm, and La) show no resolved variation in abundance ratio. Refractory lithophile abundances in CV average 1.33 times as great as those in CI and are well-resolved from the next-highest values in CM and CO chondrites. Refractory lithophile abundances in CO and CM are essentially identical, averaging 1.11 times those in CI; they are well resolved from CI abundances. Refractory-lithophile abundances in the other chondrite groups not shown in Figure VIII-3 are significantly lower than CI levels.

As discussed in Chapter II, relative to other chondrite groups, abun-

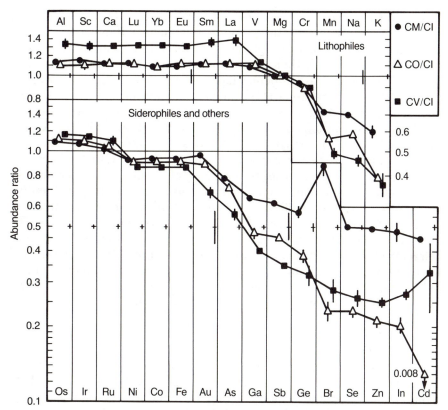

Figure VIII-3. Magnesium-normalized abundances of 30 elements in CM, CO, and CV chondrites relative to those in CI chondrites. The vertical axis is (element/Mg)$_{group}$/(element/Mg)$_{CI}$. Lithophiles (elements that condensed mainly as oxides) are in the upper portion; siderophiles (elements that condensed mainly as metals) and other elements (most of which condensed as sulfides) are shown in the lower portion of the drawing. In each portion, volatility (as measured by the nebular condensation temperature) increases to the right, except that the five rare earths lutetium through lanthanum are in the order of decreasing atomic number. The bars show 95-percent uncertainities for the group means; if no bar is shown, the uncertainty is less than the size of the symbol. The four groups are readily resolved in terms of volatiles, and only CM and CO cannot be resolved on the basis of refractories. (Data from G. W. Kallemeyn and J. T. Wasson, *Geochim. Cosmochim. Acta* 45:1217, 1981.)

dances in CI chondrites appear closest to those in the Sun and therefore probably closest to those in the primitive solar nebula. If the nebula had CI abundances of refractory lithophiles at the **formation location**[3] of the CV, CM, and CO chondrites, then these chondrites must be enriched in a refractory-rich component. It has been suggested that this component was compositionally closely related to the coarse-grained inclusions in the Allende chondrite (Figure VIII-4), which also show no interelement fractionation among refractory lithophiles. In the CV chondrites, these inclusions can be quite large (up to 2.5 cm), and it seems probable that it was because of their large size that these materials were enriched in the solids that collected at the nebula median plane prior to formation of the first planetesimals.

Two siderophiles (osmium and iridium) also are refractory; ruthenium abundance ratios are significantly lower, and ruthenium appears to be intermediate in behavior between the refractory and common elements. The CM/CI and CO/CI ratios for osmium and iridium are slightly lower than refractory-lithophile ratios, but the differences are within experimental uncertainty. Thus the data are consistent with a single component containing the refractory lithophiles and siderophiles in their solar relative proportions, and this component accounts for most of the refractories in these chondrites.

In contrast, CV/CI osmium and iridium ratios are only 1.14 times the CI ratios, whereas refractory lithophiles are 1.33 times the CI ratios. Thus, the preceding scenario cannot have held at the CV formation location; either siderophile abundances in the refractory component were lower than those of lithophiles, or an appreciable fraction of the refractory elements were in other components. This observation suggests that the distribution of elements among different minerals and the grain-size variation of these minerals varied from one formation location to the next, and that it is hazardous to assume that fractionations observed in one chondrite group occurred in identical fashion during the formation of another group. Another example of differing behaviors is noted for vanadium, which either (1) is less refractory in CV than in CO and CM, or (2) is behaving as a siderophile rather than as a lithophile, as plotted in Figure VIII-3.

The mean abundances of common siderophiles nickel, cobalt, and iron are marginally resolvable at 0.90 in CM, 0.88 in CO, and 0.85 in CV. The abundance of the common lithophile chromium in CM, CO, and CV is

---

[3] **Formation location** designates the distance from the Sun at which a particular body formed. Because the absolute distance depended on the mass of the Sun and this mass may have decreased appreciably by loss of a T-Tauri wind, the distance scale should be taken to be relevant to present locations of the planets. For example, 1 AU means the same formation location as that for the typical matter of the Earth.

Figure VIII-4. Fractured surface of the Allende CV chondrite showing light-colored refractory inclusions set in a dark-gray matrix. The two large, round inclusions show the typical shapes of the coarse-grained inclusions; fine-grained inclusions show more irregular shapes. At 2.5 cm, the larger round inclusion is the largest ever observed in a CV chondrite. Note the parallel orientation of the long axes of some of the smaller white inclusions. (Photo from R. S. Clarke et al., *Smithson. Contrib. Earth Sci.* 5:1, 1970.)

similar, possibly indicating that chromium condensed as a metal rather than an oxide. The covariation of the siderophiles implies that a large fraction of each siderophile condensed as metal. Nebular equilibrium calculations indicate that iron oxidizes out of the metal phase at temperatures of 700 ± 200 K, whereas cobalt and nickel oxidize at temperatures 100 to 200 K lower. The absence of appreciable Co/Fe or Ni/Fe fractionations implies that any such selective oxidization has left the iron in the same component (the same accreted particles) with the cobalt and nickel. Petrographic examination shows that there is very little metal (40 mg/g or less, with rare exceptions) in any of these groups today, and it also shows large Ni/Fe fractionations among minerals—for example, low Ni/Fe ratios in Fe_3O_4 (magnetite), but high Ni/Fe ratios in certain sulfides. These observations are consistent with petrographic evidence indicating that some or all of these minerals were not formed in the nebula but resulted from interactions with fluid phases in planetesimals or parent bodies.

Within each group, abundances of volatile elements show a monotonic decrease with decreasing condensation temperature; the only exceptions are a high bromine abundance in CM and slightly elevated indium and cadmium levels in CV. Volatile abundances are significantly higher in CM chondrites, which contain ~ 55 percent fine-grained matrix, than in CO and CV chondrites, which contain ~ 30 and ~ 40 percent matrix, respectively.

Especially interesting is the difference between CV and CO chondrites. All normal members of both groups are petrographic type 3. Although mean volatile abundances are essentially the same in each group, these high-precision data show that the CV/CO ratio is less than unity for all volatiles with 10^{-5} atm condensation temperatures (see Appendix G) above 650 K, and greater than unity for all volatiles with lower condensation temperatures. The cutoff temperature approximately corresponds to the condensation temperature of the major phase FeS, and a simple if oversimplified explanation of the observations is that CO chondrites accreted materials condensed as metals or oxides more efficiently than CV chondrites did, but that the accretion of sulfides was more efficient at the CV formation location.

We can take this scenario one step farther by noting that FeS forms on the surfaces of metal grains by the reaction

$$H_2S(g) + Fe(s) = FeS(s) + H_2(g)$$

where g and s stand for gas and solid phases, respectively. This reaction implies that FeS and Fe–Ni should show a strong correlation, and such a relationship has been found in bulk studies of chondrites and also in studies of separated fractions of unequilibrated chondrites. The FeS/Fe ratio should not be constant, however, but should increase with increasing surface/volume ratio and thus with decreasing grain size of the precursor metal. Thus, although the CV chondrites seem to have accreted less total metal (or its oxidized or sulfurized reaction product) than have the CO chondrites, they seem to have accreted more fine-grained metal (or its reaction products).

A general decrease in abundance with decreasing condensation temperature is commonly observed in groups of chondrites. In Figure VIII-5, group/CI abundance ratios are shown for moderately volatile elements (those having nebular condensation temperatures between those of phosphorus at the high end and zinc, tellurium, or sulfur at the low end; see also the slightly different definition in the next section) in CM and H chondrites. If the abundance of a volatile element depends only on its condensation temperature, one should observe a monotonic decrease in the abundance ratio as one proceeds to the right on these diagrams. Such a systematic trend is not observed.

There are many uncertainties involved in the calculation of condensation temperatures, which may account for a part of the discrepancy. However, a

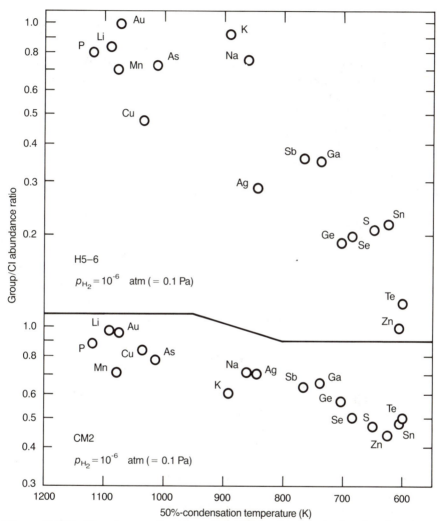

Figure VIII-5. Abundances of moderately volatile elements in the chondrite groups tend to decrease with decreasing nebular condensation temperature. H and CM data are shown as examples. This relation is interpreted to mean that, with increasing volatility, increasing fractions of the elements condense onto grains too fine to settle to the median plane of the nebula. (After C. M. Wai and J. T. Wasson, *Earth Planet. Sci. Lett.* 36:1, 1977.)

larger part probably results from the oversimplified nature of this model. Abundance ratio is a measure of accretion efficiency — the fraction of the element that condensed into or onto grains that were coarse enough to settle to the nebula median plane and agglomerate there. Two elements such as sodium and gallium may have quite different cosmochemical properties but similar condensation temperatures. If gallium condenses mainly in metal and sodium mainly in or on aluminum-rich grains, then the fractions settling to the midplane before planetesimal formation could be quite different, and they probably varied from place to place. The siting of the volatiles probably varied systematically with location; such variations may account for occasional reversals in the accretion efficiency — for example, the fact that gallium has a higher abundance ratio than sodium in CM but a lower ratio in H chondrites.

Data on the largest groups of iron meteorites were shown in Figure IV-1. For the magmatic groups, one can infer the composition of the bulk magma from the composition of the low-nickel extreme of the group and Equation IV-3 (page 81). Because the first solid crystallized in equilibrium with the unfractionated bulk magma (and with the bulk core if only one magma was present), the concentration X_I of an element X in this initial solid is related to the mean magma composition X by the solid/liquid distribution ratio k_X:

$$X_I = k_X X$$

The mean siderophile concentrations of the magmatic iron-meteorite groups have been estimated on the basis of this procedure. In Figure VIII-6, group/CI abundance ratios in IIIAB (the largest group of iron meteorites) are shown together with abundance ratios of H-group chondrites. Because the usual normalizing element, silicon, is a lithophile and thus has low concentrations in iron meteorites, the common siderophile nickel is used for normalization. As in Figure VIII-3, the elements are arranged in order of condensation temperature (decreasing to the right).

Abundance ratios of the refractory and common elements and the first three volatiles (P, Au, As) are the same in the IIIAB core and in H chondrites to within experimental error. The remaining volatiles have lower abundances in the irons than in the chondrites. The bulk of these differences for copper, gallium, and antimony seems attributable to the partitioning of the element into another portion of the body during core formation — most copper and some antimony may have entered a separate sulfur-rich magma, whereas 10 to 30 percent of the gallium may have stayed behind in the silicate mantle. Because germanium is a highly siderophile element, the 0.7-times-lower germanium value in IIIAB may indicate that slightly less germanium was present in the chondritic materials parental to IIIAB than in

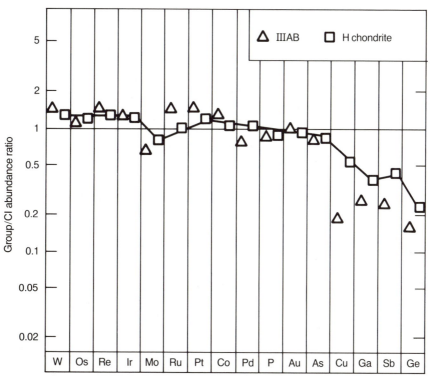

Figure VIII-6. Siderophile abundances in H chondrites and IIIAB irons are normalized to the common element nickel and divided by those in CI chondrites. The IIIAB abundances are based on the composition of the initial solid (see text for details). The iron meteorites probably formed by the melting and differentiation of chondritic bodies. The similarity in the siderophile pattern in the IIIAB and H groups implies that the IIIAB parent body consisted of materials whose composition closely resembled H chondrites. (Diagram adapted from J. Willis, Ph.D. Thesis, UCLA, 1980.)

H chondrites. In any case, this diagram illustrates the usefulness of volatile-element patterns for inferring the properties of the chondritic materials parental to the magmatic groups of iron meteorites, and it suggests that the IIIAB irons originated in the same general nebular region as did the ordinary chondrites.

Siderophile abundances are nearly constant in each chondrite group, but they vary by at least a factor of 2 between groups (for example, between EH and EL chondrites), and the range may have been larger at some nebular locations. In Figure VIII-7, silicon-normalized abundances of the volatiles germanium and gold and the refractory iridium in L and H chondrites are

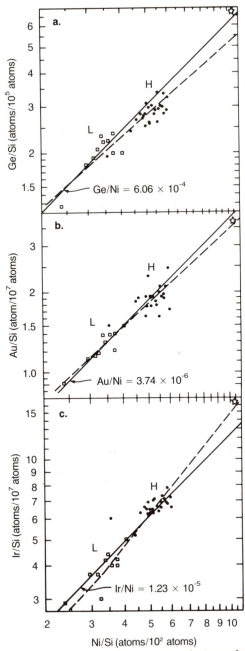

Figure VIII-7. Fractionation of siderophiles among the H and L ordinary chondrites and the ungrouped Netschaevo chondritic materials (star); data are shown as ratios to the common lithophile silicon. The solid lines show constant siderophile/nickel ratios for reference. Small fractionations seem related to volatility, which decreases Ge > Au > Ni > Ir; the Ge/Ni ratio decreases from L to H chondrites, whereas the Ir/Ni ratio increases through this sequence. Netschaevo appears to be an ordinary chondrite having siderophile abundances much higher than those in the H group. (Diagram from R. W. Bild and J. T. Wasson, *Science* 197:58, 1977.)

plotted against nickel abundance. Mean abundances in group H are higher than those in L by factors of 1.2 (for germanium) to 1.8 (for iridium). The solid line on each diagram corresponds to a constant siderophile/nickel ratio as indicated. Within each group there is considerable scatter that obscures the details of the siderophile fractionations, but it is nonetheless clear that the mean Ge/Ni ratio is higher in group L than in H, and the mean Ir/Ni ratio is higher in H than in L; the Au/Ni ratio is marginally higher in L than in H. Thus, we require a model that can explain two facts: the L group has lower siderophile abundances than does the H group, and the volatile/ refractory ratios in L-group siderophiles are higher than those in the H group.

The overall fractionation of siderophiles from lithophiles is often called a metal–silicate fractionation, and it seems reasonably probable that a partial separation of these materials did occur in the solar nebula — presumably as a result of differences in settling rates between metal-rich and silicate-rich grains. These data show, however, that the siderophiles were not confined to a single metal component having a uniform composition, but that (as usual) the situation was more complicated. We require either (1) different metal components, or (2) grain-to-grain variations in the metal composition, perhaps as a function of size (the finer the grain size, the larger the content of volatiles). There is evidence to support both views.

Experiments on size fractions of ordinary chondritic metal, on magnetic separates of whole rock samples, and on separated chondrules have independently shown Ir/Ni or Ir/Ge ratios more than 10 times higher in fine grains or in nonmagnetic silicate separates than in bulk metal. These observations are interpreted to indicate that fine, refractory metal grains condensed before the common siderophiles and became trapped in early-condensing silicates and thus isolated from further reaction with nebula gases. If the H group contained more of this refractory metal than did the L group, then the excess of iridium in the H group would be explained.

This model does not explain the details of Figure VIII-7, however. The refractory metal should have contained very little nickel and virtually no gold or germanium — yet it appears that the Ge/Au ratio is higher in L than in H. Another argument against a simple model calling for "normal" metal and refractory metal trapped in silicates is that studies of highly unequilibrated H and L chondrites show many similarities (and few differences) in the compositions and textures of the silicate component. Even though the equilibrated members of the groups differ in oxygen isotope composition (see Figure III-10 or IX-9), the most-unequilibrated show wide, overlapping fields. It appears that the silicates in these groups are from closely related nebular reservoirs, and thus it seems unlikely that the amount of trapped refractory metal in H-group silicates was appreciably higher than that in L-group silicates.

At this time, the siderophile/nickel variations seem best understood in terms of L-group metal (especially the component carrying the bulk of the common siderophiles) having a smaller mean grain size. The earliest-formed metal grains nucleated on preexisting refractory grains and attained the largest sizes; some of the later-formed grains nucleated in the gas phase and did not include any refractory metal. These later grains were generally smaller and, because they had higher surface/volume ratios, were more effective at condensing volatile elements such as germanium and gold.

The nickel and iron abundances in the H group are roughly the same as those in CI chondrites. This fact implies equal efficiency in condensing and accreting common lithophiles and siderophiles. Figure VIII-7 and Table II-1 show, however, that L-group nickel and iron abundances are only 65 to 70 percent of those in the H group; siderophile abundances in the LL group are lower yet. What happened to the missing siderophiles?

I suggested earlier that the refractory lithophiles missing from the ordinary and enstatite chondrites remained suspended as fine particles in the nebular gas and were eventually swept away together with that gas. Such a model seems less plausible for metal grains because (or so it is widely held) they are readily welded during low-speed collisions, and this coagulation leads to a gradual grain growth with time. Thus it seems implausible that 35 percent of the common siderophiles could have remained suspended in the nebular gas of the LL region at the end of the period of planetesimal formation.

An alternative is sequential formation of planetesimals systematically differing in their siderophile abundances, with the mean of the entire sequence corresponding approximately to CI siderophile abundances. Those at one end of the sequence (probably the beginning) would then have siderophile abundances significantly higher than CI levels, with those at the other extreme perhaps resembling the LL chondrites in having siderophile abundances only about 0.5 times as great as those in CI chondrites. According to this model, there should exist ordinary chondritic materials with siderophile abundances greater than those in CI chondrites. Plotted at the upper-right corners on the diagrams in Figure VIII-7 is the composition of the chondritic fragments found in the Netschaevo iron meteorite. These values seem to correspond to the anticipated siderophile-rich materials in terms of siderophile trends (only germanium is 25 percent higher than predicted by the dashed curve) and also in terms of oxygen-isotope composition (see later discussion), bulk composition, and olivine composition. The $FeO/(FeO + MgO)$ ratio in the latter is 0.14, which follows the decreasing trend of this ratio through the sequence $LL \rightarrow L \rightarrow H \rightarrow$ Netschaevo.

The systematic change in oxidation state and its relationship to siderophile abundances has been discussed for many years. Some of the relevant evidence is presented in Figure VIII-8; bulk Ni/Si in ordinary chondrites is

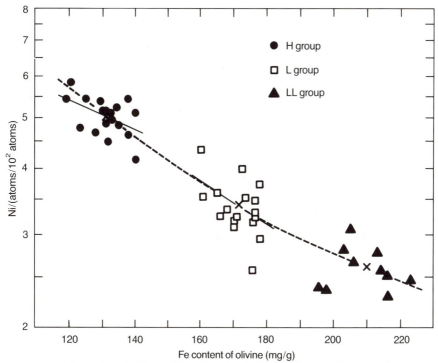

Figure VIII-8. A plot of siderophile abundance (the Ni/Si ratio) versus degree of oxidation (the iron content of olivine) is expected to follow the dashed line if the three ordinary chondrite groups are segments of an incompletely sampled, continuous fractionation sequence. Despite a large amount of experimental scatter, correlation lines in the H and L do follow the expected trend, but LL data do not correlate. (Diagram from O. Müller, P. A. Baedecker, and J. T. Wasson, *Geochim. Cosmochim. Acta* 35:1121, 1971.)

plotted against the iron content of the olivine. A dashed line connecting the three groups shows the trend expected if each of the three groups represents a segment of a continuous fractionation sequence. The basic premise of this model is that chondritic materials accumulated from several silicate and metal components, and that those meteorites that accreted more metal also obtained a larger amount of a reducing agent (or a smaller amount of an oxidizing agent) that, during metamorphism, resulted in less FeO in the primitive silicates of the more metal-rich materials. Although there is much experimental scatter in the Ni/Si ratio, the H and L data are correlated and give regression slopes (shown by the solid lines) roughly the same as those expected from the continuous-fractionation-sequence model. Because of severe sampling variations, the LL data are not correlated. Studies of unequilibrated ordinary chondrites show that several reducing agents such as elemental silicon, carbon, and phosphorus are carried by nebular metal; the silicates may have contained carbonaceous matter as a reducing agent and some matrix magnetite as an oxidizing agent.

The continuous-fractionation-sequence model thus implies that the different kinds of ordinary chondrites could have formed sequentially at the same nebular location. The required processes are in accord with those believed to be associated with planetesimal formation. There are, however, some unresolved problems with this interpretation. The three groups do not seem to have originated in the same parent body, because the cosmic-ray and metamorphism-age data discussed in Chapter III show significant differences in age clusters. Specifically, the L-chondrite body suffered one or more collisions ~ 500 Myr ago that caused extensive metamorphic resetting of ages; a far-smaller fraction of the H and LL chondrites show evidence of such recent metamorphic reheating. The H chondrites show a large cosmic-ray age cluster at ~ 6 Myr, implying a major collisional event at that time; there is no cluster in L or LL ages at that time.

Another observation is that, although many ordinary chondrites are breccias and although a sizable fraction of these contain foreign materials (mostly CM-chondrite-like materials), it is extremely rare that a fragment of one ordinary-chondrite group is found in a breccia belonging to another group. The only cases in which the foreign fragment has dimensions greater than 0.5 cm are the St. Mesmin LL chondrite (which contains an H-group fragment ~ 3 cm long) and the Dimmitt H chondrite (which contains a 1.4-cm LL fragment). If H, L, and LL planetesimals or even small asteroids formed at exactly the same distance from the Sun, it seems probable that their breccias would show far more fragments of the other two groups than of CM materials that probably (see Chapter IX) formed at a much-greater distance from the Sun.

At least equally definitive are the differences in oxygen-isotope compositions among the three groups. These differences in $\delta^{17}O$ (see Table II-1) are difficult to reconcile with a model calling for formation at the same distance from the Sun, although selective oxygen loss during metamorphism and reduction might possibly explain the observation.

It thus appears that a more-complex model is required to explain the siderophile-oxidation-state fractionations among the ordinary chondrites. At the start of this chapter, I noted that radial transport of solids relative to gas is also expected. I won't attempt to fill in the details of the picture, but it appears that such a fractionation process is required to fit the breccia and age evidence. Radial transport of metal relative to silicate components may have produced a continuously fractionated dust layer as a function of distance from the Sun, at least in the restricted regions where the ordinary- and the enstatite-clan chondrites formed. In this case, all the planetesimals at each radial distance were the same within narrow limits, and the H-, L-, and LL-group planetesimals formed far enough apart to make cross-contamination during parent-body accumulation a relatively rare event.

Fractionations Within Groups and Within Meteorites; Components in the Solar Nebula

The existence of nebular components has been mentioned several times in this chapter. For example, there is no doubt that there was a distinguishable refractory component at the CV formation location, because many of the white inclusions readily visible in Figure VIII-4 have been shown to consist of nearly pure refractory materials. Most nebular components were finely divided, however, and are not generally present as obvious grains or grain assemblages either in hand specimen or thin section. Their presence can in some cases be inferred on the basis of variations in abundance ratios between groups; an example is the variation in refractories between the carbonaceous-chondrite groups (see Figure VIII-3). Nebular conditions varied systematically with distance from the Sun, however, and it is probable that these variations were accompanied by significant variations in the composition of nebular components, including the refractory component.

In order to infer the chemical and physical processes occurring throughout the nebula, we must characterize the components that formed at each location. The ideal way would be to study samples of these components isolated from highly unequilibrated meteorites. However, many of these components had very-small grain sizes and are thus (1) difficult to isolate from other fine-grained materials, and (2) very susceptible to alteration, even during mild metamorphism. A volatile component may exist only as thin deposits on the surfaces of a nonvolatile grains formed at higher temperatures and thus be impossible to separate by mechanical techniques. In some cases, we can measure several key properties of the components by sophisticated techniques such as scanning or transmission electron microscopy, but these techniques offer limited compositional information.

Thus we commonly are unable to study an isolated component. An alternative is to study materials that consist of mixtures of components, and to infer the natures of these components from the observed compositional variations. If these materials are relatively coarse, then they may have escaped significant alteration during the minimal amount of metamorphism experienced by the highly unequilibrated chondrites.

An example of a material that appears to be coarse enough to have behaved as a closed system is a chondrule in an unequilibrated ordinary chondrite. Many of these are in the size range of 0.5 to 2.0 mm. Metal is more sensitive to metamorphic equilibration than are silicates, and studies of the chondrule metal in the most-unequilibrated ordinary chondrites show wide compositional ranges indicating disequilibrium, and thus indicate minor to negligible postnebular metamorphic alteration.

Studies of about 30 elements as well as some petrological properties (such as olivine, pyroxene, metal, and FeS abundance) in chondrules separated from the highly unequilibrated chondrite Semarkona show a wide range of compositions. A few of these deviations may be associated with volatilization during chondrule formation, but most appear to reflect the incorporation of variable amounts of several nebular components into the individual chondrules when these were formed by melting more-or-less random mixtures of these components.

Because of the large quantity of data, we must use sophisticated statistical techniques to show interrelationships. One useful technique called factor analysis links correlated properties together into "factors." Factor analysis is based on the assumption that variations in a large number of variables are due to a small number of uncorrelated factors. The factors are defined such that they can explain a maximum amount of the variation in the examined properties. Because the chondrule data set consists mainly of elemental concentrations, the derived factors can be associated with components in the precursor materials.

Figure VIII-9 shows "factor loadings" for several elements in 35 large (0.5 to 1.5 mm) chondrules from the Semarkona chondrite. For each point representing an element, the distance from the center is the square root of the fraction of its variation (called the variance) accounted for by two factors. The inner circle at radius 0.5 marks 25 percent of the variance, the outer circle 100 percent of the variance. If the point lies at radius less than 0.3, no point is plotted because the factors do not explain a significant amount of the variance. The amount contributed by each factor is obtained by drawing a line through the point perpendicular to the line from the center towards the factor and reading the intercept point. For example, the factors are able to account for about 25 percent of the bromine variance; the F2 factor alone accounts for most of this.

The F2 factor accounts for ~80 percent of the variance in five siderophiles and in the chalcophile selenium that is mainly in FeS and therefore is associated with a metal–troilite nebular component. It also accounts for surprisingly little of the variance in iron — only about 10 percent. Apparently, much iron was not in the metal component when the Semarkona chondrules formed, whereas (as discussed for the carbonaceous-chondrite groups) the coherency of iron, nickel, and cobalt in whole-rock studies suggests that these elements were in the same component. Perhaps this observation reflects metal loss by Semarkona chondrules during the period they were molten. An earlier factor-analysis study of Chainpur chondrules showed iron loading on the metal component. I cannot explain why Semarkona and Chainpur behave differently from each other.

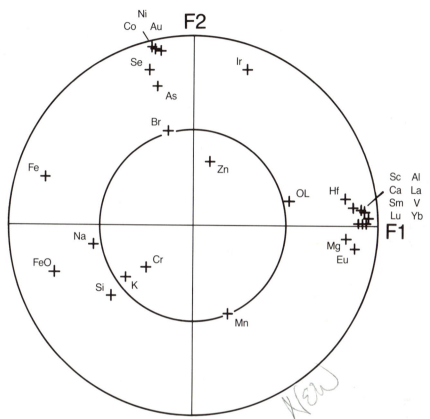

Figure VIII-9. Interrelationships of compositional data for 35 chondrules from the Semarkona chondrite were evaluated by a statistical technique called factor analysis. Elements that are closely related are gathered together as factors; in this diagram, the members of a pair of factors are plotted against each other. The distance from the center of the circle is a measure of the fraction of the total variation that is explained by either or both factors. Factors can be associated with specific nebular phases on the basis of the elements associated with them; for example, factor 2 seems to consist of metal and troilite (FeS) (see text). (Diagram from J. Grossman and J. Wasson, in *Chondrules and Their Origins*, ed. E. King, Lunar Planet. Inst., 1983.)

The F1 factor accounts for at least 85 percent of the variance in five refractory lithophiles and in magnesium and vanadium, which are intermediate in volatility between silicon and refractories such as calcium. This factor is therefore associated with a refractory-lithophile nebular component. About 80 percent of the variance in magnesium is explained by this component. Investigations of refractory inclusions and equilibrium calculations both show that only a minor fraction of magnesium condenses with the refractories. The apparent explanation of the refractory–magnesium co-variation is that the refractory component did not consist of pure refractories, but that Mg_2SiO_4 (magnesian olivine, the next silicate to condense) was mixed intimately with the refractory minerals, perhaps because it formed coatings on them. This hypothesis is supported by the observation that the olivine/(olivine + pyroxene) ratio indicated by OL in Figure VIII-9 loads at more than 50 percent on the refractory factor.

In the ordinary chondrites, two categories of volatiles can be distinguished on the basis of their relationship to metamorphic recrystallization. As discussed in Chapter II, the ordinary chondrites are assigned to petrographic types 3 through 6, largely on the basis of the extent of preservation or loss of features such as glass and chondrule boundaries associated with the original primitive texture. Abundances of moderately volatile elements show negligible variation as a function of petrographic type (type-3/type-6 concentration ratios not greater than 2); highly volatile elements tend to decrease with increasing petrographic type (3/6 ratio greater than 2).

Variations in abundance of two highly volatile elements, xenon and indium, with petrographic type are illustrated in Figure VIII-10. The highest concentrations are in L3, intermediate concentrations in L4, and lowest concentrations in L5 and L6 chondrites. Although this diagram shows L5 indium concentrations systematically higher than those in L6 chondrites, a larger data set shows extensive overlap of indium concentrations in these two types. Similar variations among the petrographic types of ordinary chondrites are also observed for planetary [36]Ar, bismuth, and thallium.

Two general models have been proposed to explain these variations in the concentrations of highly volatile elements.

1. The different chondrite types were produced in sequential solar-nebula processes. The material that condensed out of the nebula first at high temperatures had low volatile contents; it accreted to the centers of parent bodies. Later materials having monotonically increasing volatile contents accreted to form successive shells on the parent body. The parent body was then heated by an internal heat source that caused increasing degrees of metamorphic recrystallization toward the center of the body but did not lead to appreciable volatile loss or transport.

2. The chondritic materials formed by nebular processs had more-or-less-uniform concentrations of the highly volatile elements, similar to those in type-3 chondrites. During the heating that produced metamorphic recrystallization, the highly volatile elements were vaporized and lost through outgassing of the planetesimals or parent bodies.

The chief advantage of the metamorphism model is that a single process accounts for both the textural and compositional variations. The chief advantage of the condensation model is that the elements showing the highly

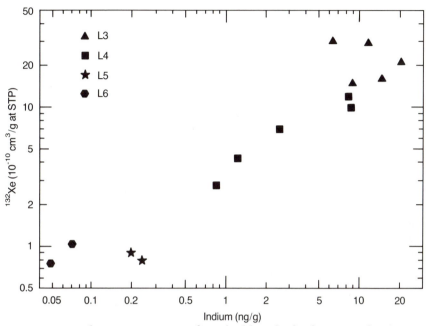

Figure VIII-10. The concentrations of two highly volatile elements, planetary
^{132}Xe and indium, are strongly correlated with petrographic type of L-group
chondrites. Concentrations decrease through the sequence L3 > L4 > L5 ≥ L6.
The number of convex corners on each symbol indicates the type number. (Data
from S. N. Tandon and J. T. Wasson, *Geochim. Cosmochim. Acta* 32:1087, 1968.)

volatile behavior are precisely those having 10^{-4}-atm nebula-condensation
temperatures below 500 K. Similar calculations are not available for the
interior or regolith of an asteroid, but it is possible that these conditions
would cause a different set of elements to be volatile. The metamorphism
model is criticized because, if the overburden pressure is great enough, the
amount of outgassing may be inadequate to produce the observed volatile
depletions. This criticism is avoided if the metamorphism and outgassing
occur in a planetesimal or in the regolith of an asteroid.

Perhaps the chief problem with the condensation model is its necessity
for synchronized timing between condensation and parent-body growth. If
the first step in parent-body growth is the formation of planetesimals, and if
this step is followed by a gradual sweeping-up of planetesimals to form
bodies of increasing size, it is remarkable that the rate of planetesimal
accretion to the final parent body was so well coordinated with the precursor steps of condensation, grain settling, and planetesimal formation that
were responsible for establishing the volatile contents of the chondritic

matter (refer to Table VIII-1). It seems much more likely that asteroids formed not by the sequential layering of successive condensates (or, more precisely, grains successively equilibrated with nebular gases) but through the parallel growth of a series of small bodies that eventually collided and merged into a single body. Even if some of the earliest bodies showed an onion-skin configuration, it is very doubtful that the final body even remotely preserved this degree of order. My current assessment is that volatile loss occurred during metamorphic heating, and that the energy source probably was impact heating of a deep regolith.

Summary of Nebular Fractionations Recorded in Meteorites

As just discussed, the composition of a chondrite reflects the combination of chemical processes that produced the set of solid components at that nebular location and mechanical processes that led to the incorporation of variable amounts of these components into the chondrite.

The scenarios presented here are only sketches of what really occurred during the formation of solid bodies in the solar nebula; the actual set of processes certainly was much more complex. Nonetheless, we know that certain processes did occur, and progress in assessing the relative importance of these processes can be achieved only by such attempts to merge meteorite research and astrophysics.

A possible criticism of this chapter is that the various models require higher initial temperatures than those predicted by most astrophysical models. My response would be that the consistent fashion in which these models account for the observed fractionation trends is itself an argument for such high temperatures. To date, no researchers have formulated a general model to account for the observed properties of chondrites in a low-temperature nebula having a maximum temperature of 1000 K at 1 AU. I do not believe that such a general model is possible.

Because a low-temperature nebula does not permit the interstellar grains to evaporate, it necessitates models that account for the fractionations described in the previous section in some fashion other than a distribution of elements into components of the sort described here. Perhaps the variations in chondritic bulk composition could be attributed to regional variations in the composition of the interstellar matter falling into the nebula. Some trends occur on a very small scale, however, such as the siderophile – lithophile fractionations among the ordinary chondrite groups. The great similarities in petrographic character and in chemical and isotopic composition seem to require that these groups were formed at about the same time and very near to each other (surely within less than 0.5 AU of each other, probably much closer—see Chapter IX), and it seems most implausible that the parental interstellar materials could have shown such large differences in siderophile/lithophile ratio over this small distance.

The presence of precursor nebular components is now well demonstrated, and these components must play an important role in all future nebula models. I pointed out that there is no plausible way for all such components to consist of small (much less than 0.1 mm) particles because (1) random-mixing processes should lead to chondrules or other chondrite components having the same compositions, and (2) relatively large sizes of refractory components are required so that their compositions remain relatively unaltered at the low nebular temperatures at which agglomeration occurred.

Despite my conviction that properties of chondrites require a high-temperature nebula, it is clear that each person tends to take a myopic view of his or her field, and it is surely good that some researchers are examining models involving low-temperature nebulas.

An important goal for the future is to gather many more isotopic data on individual chondritic components (chondrules, different kinds of extrachondrule materials and so on), and to attempt to use these data to refine both the chemical constraints and (with age dating) to determine the duration of sequencing of certain nebular processes. Examples are the determination of precise oxygen-isotope compositions and $^{129}I - ^{129}Xe$ ages of individual chondrules from highly unequilibrated ordinary chondrites. Both kinds of studies are now in progress.

Another (more difficult) goal is to construct complex nebular models that attempt to account for the detailed fractionations as a function of place and time. For example, do the similar siderophile–lithophile fractinations among ordinary-clan chondrites and among enstatite-clan chondrites indicate local nebular structures such as turbulence cells?

Suggested Reading

Anders, E. 1971. Meteorites and the early solar system. *Ann. Rev. Astron. Astrophys.* 9:1. A technical account of one view of nebular fractionation.

Grossman, J. N., and J. T. Wasson. 1983. Refractory precursor components of Semarkona chondrules and the fractionation of refractory elements among chondrites. *Geochim. Cosmochim. Acta* 47:759–771. A detailed technical review of chemical variations among chondrules from the Semarkona meteorite and implications for the nebular fractionations of refractory elements.

Larimer, J. W. 1978. Meteorites: Relics from the early solar system. In *The Origin of the Solar System*, ed. S. F. Dermott, p. 347. Wiley. A technical account of a model to account for the chemical fractionations observed in the chondrites; the view are closely related to those in the Anders (1971) review.

Wasson, J. T. 1974. *Meteorites.* Springer. Chapter 18 reviews the nebular processes that may have been involved in the formation of ordinary chondrites.

Wood, J. A. 1979. *The Solar System.* Prentice-Hall. Chapter 7 discusses nebular processes and the origin of the planets.

CHAPTER IX

Relationship to the Planets, Asteroids, and Comets: At What Solar Distances Did the Meteorites Form?

One of the most important goals of meteorite research is to understand how the planets formed, especially the rocky terrestrial planets. It appears that chondritic planetesimals formed directly in the solar nebula, and that the planets formed by the accumulation of these planetesimals, of protoplanetary bodies of intermediate size, and probably also of fine chondritic dust. An alternative proposed by some researchers is that planets formed mainly by the direct accretion of the components (such as chondrules, metal grains, refractory inclusions) that one can obtain by picking apart chondrites — in other words, that bodies of intermediate size did not play a major role in planet formation. Because meteorites consisting exclusively of any one component never fall, it seems more realistic to limit our models to chondrites, or materials having chondritic bulk compositions. Each planet, however, has different properties (even after correction for effects that depend on size), implying that the mean properties of the planetesimal building blocks varied with distance from the Sun.

It is very difficult to determine the bulk composition of a planet, even when you live on it. We do not know key elemental ratios such as Fe/Si, Ca/Si, and Al/Si in the bulk Earth to accuracies better than ±20 percent. The problem is compounded many times over when information must be obtained remotely. Although we now know far more about the planets from Mercury through Saturn than we did before these were visited by spacecraft, we have appreciable compositional data only for the Mars and Venus atmosphere and for soil samples at two Martian and five Venusian locations.

As discussed in more detail later in this chapter, there are reasons for believing that the meteorites formed over a wide range of distances from the Sun. Their formation locations probably spanned the range from the Mercury–Venus region (less than 0.7 AU from the Sun) to the outer edge of the asteroid belt (3.5 AU), and the outer limit on the range may have been beyond Jupiter. Although we will never be able to pinpoint the formation location of most meteorites, it is important to try to define the distances of their formation locations from the Sun even to within ∼20 percent, because this is the key step required to determine which kinds of meteoritic materials are the main components of each of the terrestrial planets.

Properties of the Terrestrial Planets

The orbital properties, masses, and sizes of the planets are summarized in Table IX-1. Pluto is included in this list for completeness but, because of its small mass (about 5 percent that of Mercury), it should no longer be treated as a planet but should be reclassified as an asteroid–comet or "stray" body.

The eight planets can be divided into two distinct groups: the four inner, or terrestrial, planets — Mercury, Venus, Earth, and Mars — are rocky objects having low volatile contents and high densities; the four outer, giant, or jovian planets — Jupiter, Saturn, Uranus, and Neptune — consist mainly of the light elements hydrogen, helium, carbon, and oxygen and therefore have low densities. Uranus and Neptune appear to have accreted all compounds that could condense from the nebula at temperatures of ∼ 50 K or above, including the bulk of the abundant light elements carbon, nitrogen, and oxygen. This abundance contrasts with the abundances of these three elements on the terrestrial planets, which are a small fraction of the solar values. Jupiter and Saturn not only contain solar abundances of carbon, nitrogen, and oxygen; they also have approximately solar abundances of hydrogen and helium, the two most common elements (see Appendix D). These two elements would not have condensed at the lowest temperatures experienced by the nebula, so they must have been collected as gases by the strong gravitational fields of protoJupiter and protoSaturn. It is thus necessary that Jupiter and Saturn formed before the nebular gases were dissipated. This is not a requirement for any of the other planets, and the absence of massive hydrogen and helium atmospheres on Uranus and Neptune indicates that most of the nebular gas was gone by the time they had sizes (a few times $M_\oplus$) such that their gravitational fields could collect it.

The densities of the inner planets offer important clues to their bulk compositions; these densities are now precisely determined as a result of visits by spacecraft. To compare them with chondritic densities, however, we must correct for the effects of pressure and temperature on the planetary densities. Planetary densities recalculated to 1 atm and 298 K are listed in Table IX-1 and shown in Figure IX-1. Unfortunately, the correction procedure introduces an appreciable uncertainty, because (1) the planets' mineralogical compositions are poorly defined, and (2) the empirical equations relating mineral density to pressure and temperature have appreciable uncertainties that increase with pressure and are ± 10 to ± 20 percent at the megatmosphere pressures present in the cores of Venus and Earth. The latter effect is responsible for the assignment of uncertainties to the corrected densities of Venus and Earth larger than those for Mercury and Mars.

Also shown in Figure IX-1 are the densities of the Moon and of three asteroids: Vesta, Pallas, and Ceres. The determination of asteroid densities is subject to many difficulties; the uncertainties are ± 5 to ± 10 percent in

Table IX-1
Orbit, mass, and size data for the Sun and planets

Body	Orbital Semimajor Axis (cm)	(AU)	Mass (g)	($M_\oplus$)	Equatorial radius (cm)	Density (g cm^{-3}) Actual	At 1 atm, 298 K
Sun	—	—	1.99×10^{33}	3.33×10^{5}	6.96×10^{10}	1.41	—
Mercury	5.79×10^{12}	0.387	3.30×10^{26}	0.0558	2.44×10^{8}	5.42	5.2
Venus	1.08×10^{13}	0.723	4.87×10^{27}	0.815	6.05×10^{8}	5.25	4.0
Earth	1.50×10^{13}	$\equiv 1.000$	5.98×10^{27}	$\equiv 1.000$	6.38×10^{8}	5.52	4.0
Moon	3.84×10^{10}	—	7.35×10^{25}	0.0123	1.74×10^{8}	3.34	3.4
Mars	2.28×10^{13}	1.524	6.42×10^{26}	0.107	3.40×10^{8}	3.94	3.7
(Asteroids)	$\sim 4.2 \times 10^{13}$	~2.8	$\sim 4 \times 10^{24}$	0.0007	—	—	—
Jupiter	7.78×10^{13}	5.204	1.90×10^{30}	318	7.14×10^{9}	1.31	—
Saturn	1.43×10^{14}	9.55	5.69×10^{29}	95.1	6.03×10^{9}	0.69	—
Uranus	2.88×10^{14}	19.21	8.70×10^{28}	14.5	2.54×10^{9}	1.2	—
Neptune	4.50×10^{14}	30.11	1.03×10^{29}	17.2	2.43×10^{9}	1.7	—
Pluto	5.99×10^{14}	39.44	1.31×10^{25}	0.0022	$\sim 1.5 \times 10^{8}$	~0.9	—

Source: Data mainly from J. R. Beatty, B. O'Leary, and A. Chaiken, *The New Solar System*, 2nd ed, Sky Pub. Corp., 1982.

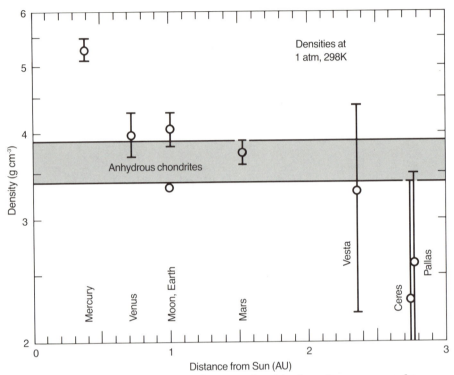

Figure IX-1. On this diagram, the planetary densities have been corrected to a pressure of 1 atm and a temperature of 298 K, which permits comparison with chondrite densities measured in the laboratory. Only Mars has a corrected density in the range of the groups of anhydrous chondrites, but the uncertainty limits for Venus, Earth, and the three asteroids are partially in the chondrite range. Mercury has a very high density, indicating that its iron abundance is much greater than chondritic values.

their masses and, more important, ± 10 to ± 15 percent in their radii. Ceres and Pallas are the two largest asteroids; Vesta is much smaller but is relatively well studied because it has a very high **albedo,** is nearer the Earth than are the larger asteroids, and by chance has had a series of close encounters that produced measurable perturbations in the orbit of Arete, a much smaller asteroid. (The **albedo** is the fraction of the incident energy, in all wavelengths, reflected by an object in space.)

A stippled band across Figure IX-1 indicates the range of densities found in the chondrite groups (EH, EL, IAB, H, L, LL, CV, CO) that do not contain hydrated silicates. Only for Mars is the best-estimate density within this range, but the uncertainties for all other bodies with the exception of Mercury overlap the range. The best-estimate densities of Ceres and Pallas are about the same as those observed in the CM chondrites. The high density of Mercury is well outside the chondritic range; this indicates that Mercury has an Fe/Si ratio much higher than that in any chondrite, because iron is the only common element with an atomic density much greater than 3 g $\cdot$ cm^{-3}. Metallic iron has a density of 7.9 g $\cdot$ cm^{-3}.

The parameters best suited to reveal relationships between the planets and the chondrites are chemical and isotopic compositions. Unfortunately, detailed information is available only for the Earth and the Moon, and precise data exist for only some of the key properties. The following are among the properties that provide the best genetic clues.

1. The Fe/Si ratio — a measure of the fractionation of metallic iron from silicates — that is, from lithophile elements.
2. The Al/Si or Ca/Si ratio — measures of the fractionation of refractory elements from common elements during nebular condensation and agglomeration.
3. The FeO$_x$/(FeO$_x$ + MgO) ratio in the silicates — a measure of the degree of oxidation.
4. The FeO/MnO ratio in the silicates — primarily a measure of the degree of planetary oxidation, but manganese is more volatile than the common elements and more chalcophilic than silicon or magnesium, and the possibility of its loss or extraction into the core must be kept in mind.
5. The K/U ratio — this ratio of a moderately volatile element to a refractory element is normally used as a measure of the abundance of moderately volatile elements; because both elements are highly incompatible, igneous processes tend to bring them into the outer crust of a planet with only a moderate fractionation in the ratio.
6. The relative abundances of the three oxygen isotopes — an empirical link to individual groups of meteorites.

Table IX-2 lists estimated values of these ratios for the terrestrial planets.

Mercury is the only planet that has no detectable atmosphere; this fact reflects a combination of (1) its low escape velocity, (2) the effective stripping action resulting from the high flux of solar photons and solar-wind protons at its location 0.4 AU from the Sun, and (3) its weak magnetic field, which is unable to prevent the solar wind from sweeping atmospheric gases from its surface. Mercury was visited by the Mariner 10 spacecraft but, although striking photographs of ~ 35 percent of its highly cratered, lava-covered surface were transmitted, no compositional data were produced. Our only direct compositional information is based on a telescopic reflectance spectrum, which is interpreted to indicate the presence of some FeO in pyroxene and the presence of minor amounts of a transition element such

Table IX-2
Key atomic or molar ratios useful in establishing relations among the planets and between planets and meteorite groups

Planet	Fe/Si°	Refractory/Si°	FeO_x†		FeO/MnO	K/U°	$\delta^{18}O$ (‰)	$\Delta^{17}O$ (‰)
			$FeO_x + MgO$					
Mercury	~5.0 ± 0.1‡	—	≤0.3§ (~0.0)		—	—	—	—
Venus	1.0 ± 0.2	—	~0.3 (<0.3)		50 ± 30	~0.04	—	—
Earth	1.0 ± 0.2	0.8 ± 0.3	~0.09 (~0.05)		60 ± 10	~0.04	5.3	0.0
Moon	0.2 ± 0.1	0.8 ± 0.4	~0.09 (~0.05)		85 ± 20	~0.006	5.2	0.0
Mars	1.0 ± 0.3	—	~0.5‖		≥8	~0.04#	—	—

° Normalized to CI-chondrite atomic ratios.
† The first value is an uncorrected value: outer-mantle values for Earth and Moon, surficial-soil values for Mercury, Venus, and Mars. The values in parentheses are roughly corrected for igneous fractionations and late-accreted chondritic matter and are thought to be more-accurate estimates of whole-planet ratios.
‡ Estimate based on EH-chondrite model with dominant phases enstatite, metal, and troilite.
§ Based on spectral analogy with the soil at the Apollo-16 lunar site.
‖ Ratio in the surface soils analyzed by the Viking landers.
The K/U ratio estimated by gamma-ray spectrometer on the Mars 1 spacecraft has a high uncertainty.

as iron in two oxidation states. The inferred amount of FeO is low and highly uncertain; it may be that the gradual accumulation in the regolith of oxidized chondritic materials (as has occurred on the Moon, as indicated by studies of lunar soils) is responsible for most of the inferred FeO. The manner in which such spectra are interpreted is discussed in the next section.

Several spacecraft have visited Venus; five, all launched by the USSR, landed and transmitted compositional data. Venera 8, 9, and 10 landed on its surface and measured the gamma-ray flux for about 1 hr each before succumbing to the high surface temperatures (~ 750 K). Venera 13 and 14 analyzed soil samples by X-ray fluorescence. The gamma-ray data from Venera 8 indicate concentrations of 40 mg/g K, 2 μg/g U, and 6.5 μg/g Th—similar to those in granitic terrestrial rocks. Venera 9 and 10 found 3 to 5 mg/g K, 0.5 to 0.6 μg/g U, and 0.7 to 3.6 μg/g Th—similar to concentrations in terrestrial basalts. The K/U ratio in Venus is inferred to be about the same as that in the Earth. Venera 13 and 14 determined eight elements including magnesium, iron, and manganese; the uncertainty in manganese concentration, and thus in the FeO/MnO ratio, was ± 50 percent. At the Venera 13 site, the composition is consistent with bedrock consisting of an alkali basalt; that at the Venera 14 site is consistent with a tholeiitic basalt.

The main Venusian atmospheric constituent is CO_2; the surface pressure is 90 atm, and 97 percent of the gas is CO_2. This is $\sim 2.5 \times 10^5$ more CO_2 than is present in the Earth's atmosphere, but the amounts on the planets are similar if one includes the CO_2 bound as $CaCO_3$ in terrestrial sedimentary rocks. Carbonates are unstable at the surface of Venus. Venus is covered with clouds that consist of H_2SO_4–H_2O droplets. An interesting discovery made by the Pioneer Venus probe is that the $^{36}Ar/^{40}Ar$ ratio in the Venus atmosphere is 0.85, whereas it is 0.003 in the terrestrial atmosphere; the ^{36}Ar partial pressure is 100 times higher on Venus than on the Earth. The rare-gas elemental ratios in the Venus atmosphere are less fractionated relative to solar values than those in the Earth's atmosphere, in most carbonaceous chondrites, or in ordinary chondrites, and they are remarkably similar to those in some enstatite chondrites.

The mantle of a planet represents a much-larger mass fraction than does the crust, and some of the most-important information about the Earth comes from mantle samples brought to the surface explosively or carried to the surface by rapidly moving streams of alkali basalts. Because we can measure important properties such a silicate $FeO_x/(FeO_x + MgO)$ or FeO/MnO ratios or oxygen-isotope compositions directly in these mantle materials, our whole-Earth estimates of these ratios are relatively precise.

Basalt is produced in the upper mantle by partial melting followed by transport of the melt liquid to the surface. As a result, some and possibly most upper-mantle materials have lost an appreciable portion of the elements that tend to concentrate in such a low-melting fraction. If we are to use upper-mantle rocks to estimate whole-Earth properties, we must either focus our attention on the most-undepleted mantle rocks or use data on crustal rocks to correct for losses occurring during partial melting. In fact, both techniques are used. One must also make the plausible but unproven assumption that the entire mantle is essentially identical in composition to the upper mantle plus most or all of the crust. We also find that we need to correct some ratios for the accumulation of oxidized chondritic materials after formation of the Earth's core. Our estimates of mean terrestrial values for the key parameters are given in Table IX-2.

The $FeO_x/(FeO_x + MgO)$ ratio in mantle rocks is tightly controlled to be 9 to 11 mol%. Allowances for the iron and magnesium in the crust leads to a negligible increase in this ratio. However, allowance for the complete oxidation of iron in chondritic material accreted to the Earth following core formation indicates that this value should be corrected downward by about a factor of 2. Similarly, the range on oxygen-isotope composition of mantle rocks is relatively narrow, ranging from $\delta^{18}O$ values of $+5.0‰$ to $+5.5‰$. The few samples in which $\delta^{17}O$ has been determined have not deviated from the expected "terrestrial" values. Recent studies of $^{87}Rb - {}^{87}Sr$ and $^{147}Sm - {}^{143}Nd$ systematics in young crustal rocks indicate that the abundance of refractory elements in the whole Earth is probably about the same as that in ordinary chondrites. The estimation of the bulk iron content of the Earth is rendered difficult by a lack of knowledge about the composition of the core. The density of the core appears to be less than that expected from $Fe - Ni$ at core pressures, probably because the core contains light alloying elements such as carbon, silicon, or sulfur. Because there is no evidence that contradicts the simplest assumption, a CI-chondrite Fe/Si ratio, we use this as our working estimate.

The CI-normalized K/U ratio of 0.04 is typical for the higher values observed in the relatively undepleted upper mantle rocks known as lherzolites. Basaltic values are about 4 times higher, presumably because potassium has a greater tendency than uranium to enter basaltic liquids. The FeO/MnO ratio in Earth rocks is relatively constant and is much higher than that found in chondritic meteorites having similar $FeO_x/(FeO_x + MgO)$ ratios. These high values seem to require "loss" of manganese from Earth materials. The two most plausible models are (1) that manganese was volatilized in the solar nebula, during differentiation of asteroid-sized planetesimals, or from the early Earth; or (2) that manganese was present as sulfides in the primitive Earth, and the missing manganese is now in the core.

During the years after the first return to Earth of lunar samples, it was widely held that the Moon is enriched in refractory elements. However, more recent data and interpretations indicate that the observed amount of anorthosite (a mineral containing large amounts of the refractories calcium and aluminum) and the compositions of igneous rocks are both consistent with formation during extensive melting of a Moon having ordinary-chondrite abundances of refractories. These same modeling experiments indicate that a $FeO_x/(FeO_x + MgO)$ ratio in the bulk Moon similar to that in the Earth is consistent with the compositions of igneous rocks. There is no doubt that the Fe/Si ratio in the Moon is well below the chondritic value; if not, the density of the Moon would be much higher. The FeO/MnO ratio in the Moon appears to be resolvably higher than that in the Earth. It seems likely that proper correction for late-accreted chondritic materials would decrease the terrestrial FeO value more than the lunar value and would thus increase the difference; there are so many uncertainties involved in such corrections, however, that it is premature to use the FeO/MnO ratio to rule out a close genetic relationship between the mantles of the Earth and the Moon. The K/U ratio in the Moon is 6 to 7 times lower than that in the Earth; the origin of this relatively large discrepancy is not clear, but it is probably connected with outgassing of the Moon or its precursors. In this case also, it appears to be premature to use this difference to rule out a close genetic relationship between these two bodies. There is no resolvable difference in oxygen-isotope composition between the Earth and the Moon.

The Martian atmosphere has a pressure of 0.0052 atm; its three main constituents are CO_2 (95 percent), N_2 (2 to 3 percent), and Ar (1 to 2 percent). The $^{36}Ar/^{40}Ar$ ratio is about 0.003, 10 times smaller than the terrestrial ratio. This lower ratio may indicate that Mars with its weaker gravitational field was less effective than the Earth in retaining primordial rare gas. The atmosphere of Mars is too cold to permit an appreciable partial pressure of H_2O, but solid H_2O is observed as ice at the polar caps and is inferred to be present as appreciable fractions (perhaps approaching 100 mg/g) of water of hydration in regolith materials. Studies by the Soviet Mars 5 orbiting gamma-ray spectrometer indicated a potassium concentration of ~ 12 mg/g and a uranium concentration of about 0.7 μg/g, suggesting that the K/U ratio on the surface of Mars is similar to that on Earth. These values must be considered somewhat uncertain, however, because details have not been published, and the reported potassium content is ~ 5 times higher than that measured by the Viking 1 and 2 landers.

Martian soil samples were analyzed by an X-ray fluorescence spectrometer on the Viking 1 and 2 landers. Attempts to find small igneous rocks to analyze were unsuccessful; each gravel-sized piece that was examined proved to be a soil clod that crumbled during handling. The soil analyses are listed in Table IX-3, together with a rough estimate of the composition of the Earth's crust. The relative abundances of the nonvolatile major and minor elements are those expected in a basaltic rock. Even if converted to high-oxidation-state oxides (Fe to Fe_2O_3, and S to SO_3), the concentrations in the S-1 and S-3 samples total only 918 and 936 mg/g, respectively,

Table IX-3
Soil compositions on Mars at Chryse (S-1, S-2, S-3) and Utopia (U-1)

Sample	Mg	Al	Si	S	Cl	K	Ca	Ti	Fe	Rb	Sr	Y	Zr
	(mg/g)									(μg/g)			
S-1	50	30	209	31	7	<2.5	40	5	127	≤30	60	70	≤30
S-2	—	—	208	38	8	<2.5	38	5	126	—	—	—	—
S-3	52	29	205	38	9	<2.5	40	5	131	—	—	—	—
U-1	—	—	200	26	6	<2.5	36	5	142	≤30	100	50	30
Uncertainty[*]	25	9	25	5	3	—	8	2	20	—	30	30	20
Earth[†]	21	60	205	0.23	0.14	12.7	38	6.5	44		53	340	26	110

[*] These uncertainties apply to the Martian samples.
[†] The crustal composition for the Earth is based on equal parts granite and basalt (after K. K. Turekian, *Chemistry of the Earth*, Holt-Rinehart-Winston, 1972) . The estimated silicon content of the Earth's crust is 272 mg/g; these crustal data are renormalized to a value of 205 mg/g to facilitate comparison with Martian soils.

indicating that a major amount of some oxide is missing; this missing oxide is probably mainly H_2O, which is not detected by X-ray fluorescence.

The $FeO_x/(FeO_x + MgO)$ ratio in the Mars soils is about 10 percent higher than that estimated for the Earth's crust. However, Mars appears to be less evolved in an igneous sense than is the Earth and, because this ratio tends to rise during continued recycling of crustal material, it seems likely that the $FeO_x/(FeO_x + MgO)$ ratio in the Martian mantle is substantially higher (perhaps more than 2 times higher) than that in the Earth's mantle.

The concentrations of sulfur and chlorine are 50 to 150 times higher in the Martian soils than in the terrestrial crust. On Earth, these elements are common in volcanic gases; they also tend to form water-soluble minerals that are leached during weathering and transported to the oceans. It appears that volatile transport and/or leaching leads to surface enhancements on Mars. The potassium upper limit in the Martian soils is about 5 times smaller than the estimated concentration in the terrestrial crust; the upper limit on the K/Si ratio is about 2 to 3 times smaller than typical values in terrestrial basalts. As noted earlier, the potassium upper limit for the soils is about 4 to 5 times smaller than the planetwide potassium concentration from the Mars 5 spacecraft. The manganese upper limit in the Mars soils is 20 mg/g, leading to a lower limit of the FeO/MnO ratio that is too low to offer meaningful comparisons with chondritic values.

Properties of the Asteroids, Comets, and Meteors

Asteroids are interplanetary objects generally having radii in the range 1 – 1000 km and, in contrast to comets, having no detectable atmosphere. Because cometary volatiles are lost by evaporation in the inner solar system, it is possible that comets evolve into objects that are asteroids in the observational sense. Most of the known asteroids are in the **asteroid belt** located between the orbits of Mars and Jupiter. Within this region are about 2000 numbered asteroids, large ($R \gtrsim 20$ km) objects whose orbits are reasonably well defined. As illustrated later in Figure IX-7, the number of asteroids in the belt increases from low values at a semimajor axis of 2.15 AU through a high plateau and then drops to low values again beyond 3.45 AU (see Appendix H for a discussion of celestial mechanics applied to heliocentric orbits). The number-versus-distance function is interrupted by several minima called Kirkwood gaps that appear at radii where the asteroid's orbital period corresponds to certain fractions of Jupiter's period.

Belt asteroids collide with one another today, and the frequency of collisions must have been much larger early in solar-system history. Some of these collisions must have led to the disruption of the original body. However, because the separation velocities must have been small (much less than 1 km $\cdot$ s^{-1}), the surviving fragments should still be in closely related orbits today. For this reason, groups of asteroids having very similar orbits are called asteroid families (commonly called Hirayama families, after the astronomer who first recognized their existence). Recent studies have shown that, whereas some of these groups must be families formed by disruptions of large asteroids, others were not formed by breakups but are the remnants of an original population situated between or among regions depleted by dynamic processes resulting from strong gravitational interactions with Jupiter.

During the past decade, a large number of asteroids has been studied spectroscopically with Earth-based telescopes. Data for about 300 asteroids in the visible and near infrared wavelength range, 0.33 to 1.07 μm, are now available. Some of the most interesting information is in the infrared region, and the spectra of some 15 to 20 asteroids have now been extended to a wavelength of 2.5 μm; for a few asteroids, spectral data to 3.7 μm are available.

Figure IX-2 shows visible and infrared reflectance spectra for six asteroids. There are several spectral regions that convey information about mineralogical compositions. The most-diagnostic information results from electronic transitions of elements having vacancies in the d shell; generally iron is the element that produces such an absorption feature, and the absorption in pyroxene at ~ 0.85 μm tends to be stronger than the corresponding feature in olivine at ~ 1.05 μm. A pyroxene absorption is seen in

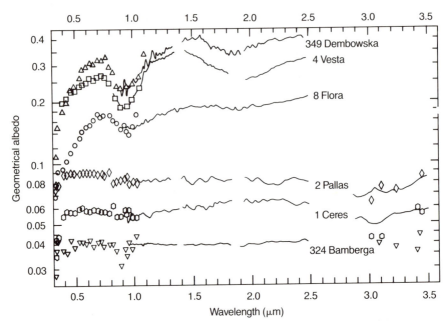

Figure IX-2. Visible and infrared reflectance spectra of six asteroids; Pyroxene absorptions at ~0.9 μm are obvious in asteroids 349 Dembowska and 4 Vesta, and minima near 1.9 μm probably are also due to pyroxene. In addition to the 0.9-μm pyroxene band, olivine absorption at ~1.1 μm appears to be present in 8 Flora. The upper three asteroids are reddish—they show gradually increasing reflectance with increasing wavelength in the visible range 0.3 to 0.8 μm; these asteroids have high albedos (greater than 10 percent). The largest asteroid (1 Ceres), the second-largest (2 Pallas), and 324 Bamberga have nearly flat spectra throughout the entire range and low albedos (less than 6 percent for Ceres and Bamberga).

the spectra of 4 Vesta and 349 Dembowska; the poorly defined minima near 1.9 μm are also attributed to pyroxene. Both pyroxene and olivine contribute to the broad absorption feature at 0.9 to 1.0 μm in the 8 Flora spectrum. The steepness of the slope below about 0.5 μm in the reflectance spectra offers information about charge-transfer transitions such as $Fe^{+2} + Fe^{+3} = Fe^{+3} + Fe^{+2}$, or $Ti^{+3} + Fe^{+3} = Ti^{+4} + Fe^{+2}$. Comparison with the spectra of rock (especially meteorite) powders shows that high slopes between 0.4 and 0.3 μm (as seen in the 1 Ceres and 324 Bamberga spectra) are found in the oxidized, carbonaceous chondrites, whereas moderate decreases (as seen in 8 Flora) are found in the spectra of less oxidized chondrites such as L-group ordinary chondrites.

Perhaps the best match between an asteroid spectrum and the spectrum typical of a meteorite class is that between 4 Vesta and the differentiated eucrites and howardites. There seems to be little doubt that the regolith on Vesta consists mainly of basaltic materials generally similar to these meteorites.

One recurring question is the solar distance at which the CM and CI carbonaceous chondrites (the two groups having hydrated layer-lattice-type silicates) originated. Such minerals are stable only at low (less than 400 K) nebular temperatures, and thus these meteorites probably formed far from the Sun. The chief competing locations are the asteroid belt (2.2 to 3.4 AU from the Sun) and the region beyond the orbit of Jupiter (more than ~5 AU from the Sun). Silicate water of hydration shows up at several infrared frequencies, but most of these also correspond to absorption frequencies of atmospheric H_2O gas, and thus they cannot be used for ground-based studies of asteroids. A broad hydration-H_2O band between 3.0 and 3.5 μm can be observed, and evidence for H_2O has been found in 1 Ceres and 2 Pallas. On the other hand several classical "carbonaceous" C-type asteroids (to be defined shortly) including 324 Bamberga do not show a resolvable absorption feature in this range. Because the H_2O absorptions in Ceres and Pallas could result from a thin veneer of cometary dust, the question of whether hydrated silicates formed within 3.4 AU of the Sun remains open.

There are several reasons why the spectra of asteroids are not readily matched by laboratory spectra measured on meteorite powders or by artificial mixtures of minerals of known composition: (1) at each wavelength, the spectrum tends to be dominated by the most strongly absorbing component; (2) the degree of absorption varies with grain size of the component (as expected because, the smaller the grain size, the greater the relative surface area); (3) some of the most-common meteoritic materials are opaques (metal, shocked olivine, carbon) having no absorption features; (4) the effects of space weathering (irradiation by the solar wind, heating and comminution by micrometeorite impacts) have not been much studied, but these probably tend to produce materials that are darker and that have less-pronounced absorptions than those found in the spectra of meteorite powders; and (5) perhaps most important, asteroidal regoliths probably consist of mixtures more heterogeneous than most meteorites, and they may contain materials different from any meteorites in our biased terrestrial sample. I discuss this biasing in more detail later in this chapter.

On the basis of their spectral reflectances, the studied asteroids have been assigned to several classes; the two largest are the C-type objects showing flat visible and near-IR spectra and S-type objects that reflect more red than blue light and typically have a pyroxene or olivine absorption band. The C-type asteroids have albedos not greater than 6.5 percent and are similar in their spectra to low-metal, carbonaceous chondrites. The S-type asteroids (which have moderate albedos of 6.5 to 23 percent) appear to consist of roughly equal portions of metal and silicates, similar to the rare differentiated meteorite groups of mesosiderites and pallasites; however, the uncertainties in interpretation are great enough to allow the amount of Fe–Ni to be as low as that in H chondrites (170 to 200 mg/g), and possibly even as low as that in L chondrites (60 to 100 mg/g). In the entire belt, C objects account for ~74 percent of asteroids having diameters greater than 50 km, S-type objects for ~16 percent, and several other minor types for the remaining ~10 percent. The fraction of S asteroids drops from ~60 percent at 2.2 AU to ~8 percent at 3.4 AU.

An albedo histogram for belt asteroids has two pronounced peaks, as shown in Figure IX-3. Those asteroids with lower albedos are mainly C-type, those with higher albedos mainly S-type. Well-defined peaks on such a histogram generally imply distinct sets of formational processes — that is, distinct origins. Possible models will be discussed later in this chapter.

Comets are mixtures of ice and stony matter recognized by their ex-

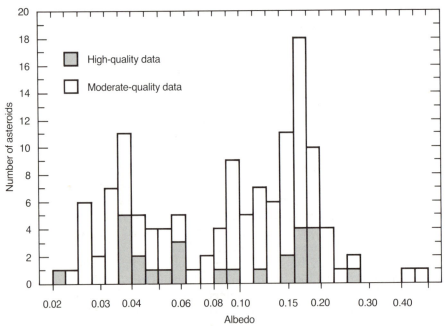

Figure IX-3. A histogram of asteroid albedos (the fraction of incident light energy reflected) shows two peaks near 0.04 and 0.17, and a minimum near 0.065. The existence of two peaks cannot be explained by simple accretionary formation of the asteroids if all formed at about the same time by processes that varied monotonically with distance from the Sun. Possible explanations of the bimodal distribution are discussed near the end of this chapter.

tended atmospheres when they enter the inner solar system. Comets can be divided into two categories: **new comets** having very large aphelia (a) and making their first pass through the inner solar system, and **periodic comets** that are in orbits having relatively short periods. Figure IX-4 shows the distribution of semimajor axes among comets. The peak at $10^4 \lesssim a \lesssim$

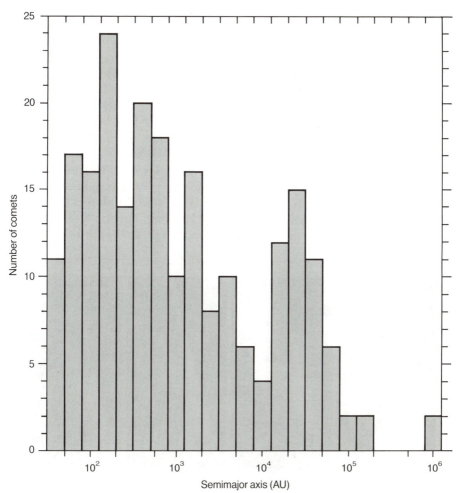

Figure IX-4. Most comets are thought to originate in the Oort cloud, a region 10^4 to 10^5 AU from the Sun that shows up as a peak in this histogram of comet semimajor axes. The estimated 10^{11} comets in the Oort cloud are occasionally perturbed by passing stars into new orbits that bring them into the inner solar system. On this diagram, the comets having semimajor axes less than $\sim 10^4$ AU have probably already made one or more passes through the planetary portion of the solar system and have had their semimajor axes reduced by gravitational interactions with the planets.

10^5 AU is interpreted to indicate that the associated aphelia (twice the semimajor axis) are measures of the mean distance from the Sun at which a cometary "cloud" is located. It is estimated that this cloud (called the Oort cloud after the astronomer who proposed this model) contains more than 10^{11} comets, but occasionally a passing star perturbs some of the orbits and causes these comets to pass into the planetary part of the solar system (within 30 AU of the Sun). While a comet is in such an orbit, perturbations by one of the planets (generally by Jupiter or Saturn) can bring the comet into still another orbit. Thus, the broad hump below 10^4 AU on Figure IX-4 consists of comets whose orbits have been altered by at least one planetary perturbation. Of course, any orbit that has a perhelion of less than 2 AU would lead to partial evaporation of the cometary ices.

The spherical atmosphere immediately around a comet is called the **coma;** typical coma dimensions are about 10^4 to 10^5 km. The atmosphere consists mainly of H_2O gas produced by the evaporation of ice. The H_2O evaporation rate increases approximately as r^{-2}, where r is the distance from the Sun. The brightness of the comet increases approximately as r^{-4}, because the amount of sunlight striking each gram of atmosphere also increases as r^{-2}. Comets have two kinds of tails, a straight **plasma tail** and a curved **dust tail;** these tails are up to 10^8 km long, but some old, nearly volatile-free comets may produce no resolvable tail. In Figure IX-5, views

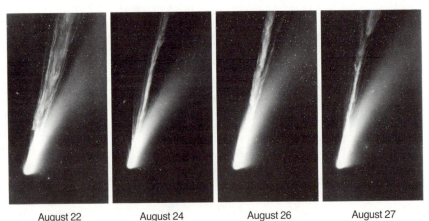

| August 22 | August 24 | August 26 | August 27 |

Figure IX-5. Photos of comet Mrkos taken on four days in August 1957 show different relative intensities of the curved dust tail on the right and the essentially straight (but ragged or braided) plasma tail on the left. The dust tail consists of particles in heliocentric orbits similar to but having radii slightly greater than that of the comet; the plasma tail consists mainly of ions that are escaping from the solar system as a result of interaction with the solar wind and its associated magnetic field. (Mount Wilson Observatory photos.)

of comet Mrkos taken during four nights in 1957 show widely varying strengths of the plasma tail (left) and dust tail (right). The plasma ions are accelerated by the magnetic field of the solar wind; thus the plasma tail points radially outward from the Sun, but it shows some kinks reflecting irregularities in the magnetic field. The interaction of small dust particles with sunlight and the solar wind gives the dust an impulse radially away from the Sun and slightly decreases the velocity parallel to the orbit. Thus these particles enter new orbits similar to that of the comet, but having slightly larger semimajor axes. As a result, their angular velocities become less than that of the nucleus, and they lag behind it to increasing degrees. The resulting dust tail appears curved because the particles continue to follow elliptical orbits.

These silicate particles eventually become distributed along the entire orbital path of the comet. When the Earth passes near these particle-laden regions (called **meteor streams**), many of the particles enter the Earth's atmosphere and fall as **meteor showers.** Tracing the atmospheric paths of these shower meteors back to the point where they first became visible shows that they all seem to radiate from the same point in the celestial sphere, the so-called radiant. This is a direct result of the fact that the orbits of each of these particles are so similar. Meteor showers are commonly named after the constellation nearest to their radiant; thus the Leonids originate in Leo, the Taurids in Taurus, and so on. Table IX-4 lists the source comets and a selection of orbital information for some of the most important meteor showers. The atmospheric velocity is the velocity at which the meteoroid enters the Earth's atmosphere.

The **cometary nucleus** is the solid part of the comet; at large distances from the Sun, it is the entire comet. A typical cometary nucleus has a radius of ~1 km, and the largest known has a radius of ~50 km. The nucleus appears to consist of icy and rocky materials in roughly equal proportions. The composition of the icy materials can be inferred by spectroscopic study of the coma and the plasma tail. The most-abundant icy constituent is H_2O; neither the exact identity nor the amount of the second-most-abundant icy constituent is known. Spectroscopic studies show several carbon-bearing species in the coma that probably formed by dissociation of an abundant ice — perhaps CO_2, CH_3CN, or a formaldehyde polymer $(HCHO)_n$.

The relative amount of the silicate materials in the nucleus of a new comet is roughly estimated to be 50 percent — that is, to be equal to the amount of ice. In an old comet that no longer has an appreciable coma, the silicate/ice ratio may be much higher. The few available compositional data on silicate particles are obtained from studies of the spectra of stream meteors in the Earth's atmosphere. These spectra show generally chondritic compositions to within the attainable relative precision of ±~40 percent. The strength and size of the meteoroids vary from comet to comet.

Table IX-4
Meteor streams, their orbital properties, and the source comets associated with them

Stream	Date[°]	Hourly rate	Atmo- spheric velocity $(km \cdot s^{-1})$	Semi- major axis (AU)	Peri- helion (AU)	Associated comet
Quandrantids	3 Jan	30	44	3.41	0.97	——
Lyrids	23 Apr	8	48	29.6	0.92	1861 I
η Aquarids	4 May	10	65	4.00	0.49	Halley
Arietids	8 Jun	40	41	1.59	0.09	——
ξ Perseids	9 Jun	30	31	1.59	0.34	——
β Taurids	30 Jun	20	33	2.22	0.34	Encke
δ Aquarids	30 Jul	15	42	2.60	0.06	——
Perseids	12 Aug	40	61	20.8	0.94	Swift–Tuttle
Draconids	10 Oct	—	26	3.33	1.00	Giaconbini– Zinner
Orionids	21 Oct	15	67	7.61	0.54	Halley
N Taurids	4 Nov	4	32	2.13	0.32	Encke
S Taurids	4 Nov	4	32	2.31	0.37	Encke
Leonids	16 Nov	6	73	12.6	0.97	Tempel
Andromedids	20 Nov	—	23	3.34	0.8	Biela
Geminids	13 Dec	50	38	1.39	0.14	1983TB[†]
Ursids	22 Dec	12	38	5.81	0.93	Tuttle

Source: After C. W. Allen, *Astrophysical Quantities*, 3rd ed., Athlone, 1973.
[°] The date on which the fall rate generally is at a maximum.
[†] An asteroid designation; this object was discovered by the IRAS infrared-astronomy satellite, and it was not initially recognized to be a comet (or former comet).

The Draconid particles produced by comet Giacobini–Zinner are remarkably fragile, as indicated by the fact that they break up very high (~ 120 km) in the atmosphere, and their luminous paths are generally complete at altitudes of ~ 100 km. The luminous portion of a typical meteor extends from ~ 110 km down to ~ 80 km, and a few larger meteoroids reach much-lower altitudes before being slowed to a nonluminous terminal velocity.

Very-large meteoroids are associated with a few showers. The Prairie Network (see Chapter I) photographed Taurid meteoroids having estimated initial masses of ~ 100 kg. The generally accepted comet model is the icy-conglomerate, or dirty-snowbank, model in which ice and dust are mixed on a scale of millimeters or centimeters. If the typical cometary silicates are large (like the 100-kg Taurid meteoroids), however, this model must be modified to allow for silicate masses having dimensions in the meter range. This modification in turn implies that the agglomeration process occurred over an extended period at the locations where comets formed, because it indicates that the silicates agglomerated before H_2O condensed from the nebula.

The large shower meteoroids photographed by the three photographic networks have never yielded recoverable meteorites, nor are any other meteorites in our collections definitely associated with a meteor shower. The meteorites occasionally observed to fall during shower periods have not had the radiant and velocity associated with that shower. This is not surprising, because most shower meteoroids enter the Earth's atmosphere at velocities that are so high that even tough meteoroids would be disrupted. Even strong stony meteorites cannot survive atmospheric deceleration as large (more than 100-g) pieces if their atmospheric velocity is greater than 25 km $\cdot$ s^{-1}. Iron meteoroids can probably survive moderately higher velocities.

As mentioned already, the strength of a meteoroid can be judged from its resistance to fragmentation during atmospheric passage. The meteoroids photographed by the Prairie Network have been assigned to three strength categories. Category-III meteoroids (10 percent of the observations) are much too weak to fall as meteorites, even when they have low atmospheric velocities; many of these meteoroids are associated with meteor streams. Category-II meteoroids (60 percent) are also weak but probably are able to survive atmospheric passage under favorable circumstances; some of these are associated with meteor streams (for example, Taurids with comet Encke). Hydrated CM and CI carbonaceous chondrites are weak, and it is possible that category-II meteoroids consist mainly of similar materials. A spectrum of one such meteoroid showed the presence of the CN radical, indicating that carbonaceous matter was present. Category-III meteoroids (30 percent) are strong rocks similar to ordinary chondrites and the other chondrites that do not contain hydrated silicates. Differentiated meteorites could be used as analogs for these meteoroid classes but are not because of their rarity.

A small fraction of the fireballs observed by the photographic networks have orbital semimajor axes greater than 5.2 AU. Dynamical calculations show that these almost certainly originated beyond 5.2 AU, and thus they can be inferred to be of cometary origin. Some of these have low atmospheric velocities and category-II strengths, and thus they have the potential to yield meteorites. Recovery of a meteorite from such an orbit may yield the first gram- to kilogram-sized meteorite of indisputable cometary origin. Because confirmation of such an association requires orbital data, it would be desirable to cover additional fractions of the Earth's highly populated regions (especially China and India) with photographic networks.

During the past decade, the first samples of **interplanetary dust** have become available for study. Interplanetary particles having dimensions less

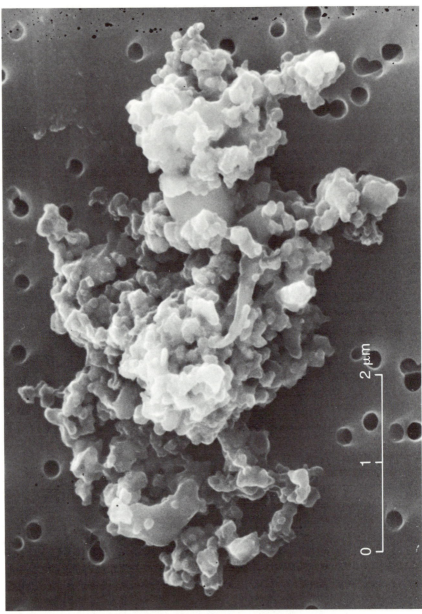

Figure IX-6. This particle was collected on a plate exposed to the air stream by an aircraft flying in the stratosphere at an altitude of 20 km. This photo was made by a scanning electron microscope; the scale bar is 1 μm. Such grains contain solar rare gas indicative of an extraterrestrial origin; they are similar to CM or CI chondrites in composition, but they differ in some properties (such as porosity). It seems likely that these materials are of cometary origin. (Photo by D. E. Brownlee.)

than 60 μm are able to survive atmospheric entry without melting because their high surface/volume ratios permit efficient heat loss by radiation. Plates exposed to the air stream of aircraft flying at high altitudes (~ 20 km) have collected small extraterrestrial particles ranging in size from 3 to 60 μm. The lower limit is imposed by the collection device; the upper limit reflects the high settling velocities and the resulting low atmospheric concentrations of particles larger than 60 μm. These stratospheric particles (often called Brownles particles after their discoverer) have compositions similar to those in CI or CM chondrites; abundances are the same to within about a factor of 1.5. Their textures (Figure IX-6) differ in detail, however, from those in CM and CI chondrites; in particular, porosities are so high that their densities are often only ~ 1 g $\cdot$ cm^{-3}. Thus, the conditions of their formation must have differed in some respects from those of the CM and CI chondrites. It seems quite likely that the Brownlee particles come from comets, but it is too early to say whether they are typical of cometary solids. Perhaps they are representative of the fine-silicate fraction of some or all comets, whereas the coarser silicate blocks in comets are more dense and are similar to CM or CI chondrites.

Orbital Variations — Then and Now

The planets are in exceedingly stable orbits, stable over periods far greater than the anticipated 10-Gyr luminous lifetime of our Sun. Most belt asteroids also are in orbits that have long dynamic lifetimes, greater than 10^{11} yr. In contrast, an object in a typical meteorite orbit with semimajor axis a of ~ 2.5 AU and perihelion of ~ 0.9 AU (just inside the Earth's orbit) has a dynamic lifetime is only ~ 10 Myr. Such objects disappear either because they impact one of the terrestrial planets (mainly the large ones, Earth or Venus) or because a gravitational perturbation during a near miss of a terrestrial planet results in a new orbit that crosses Jupiter's orbit, and the powerful gravitational field of Jupiter quickly (within ~ 100 kyr) ejects the object from the solar system.

When the planets were forming 4.5 Gyr ago, there was a large population of planetesimals in Earth-crossing orbits. If these had dynamic halflives of 10 Myr, the fraction surviving until today (after 450 halflives) would be $2^{-450} \cong 10^{-135}$, a vanishingly small number. For this reason, we can be sure that no meteorite falling today can be a remnant of an original population that occupied similar orbits at the beginning of solar-system history. Because meteorites do fall today and some asteroids (the so-called **Apollo asteroids**) occupy similar orbits that come within 1 AU of the Sun, there must be processes that bring objects into Earth-crossing orbits from "storage" orbits having dynamic lifetimes comparable to the 4.5-Gyr age of the solar system. The two chief locations where such materials could be stored are the asteroid belt and the Oort cloud of comets. Several mechanisms have been proposed for injecting these materials into Earth-crossing orbits.

The dynamic halflife of an asteroid in an orbit crossing that of Mars and having an aphelion small enough (less than 4.5 AU) to avoid major Jupiter perturbations is about 1 to 5 Gyr. With such a halflife range, 1 to 50 percent of the original population should still be present today. Objects are lost from this population largely as a result of repeated orbital perturbations by Mars that eventually reduce the perihelion to less than 1 AU, after which the much-stronger field of the Earth reduces the residual halflife to ~ 10 Myr, as discussed earlier. There is no doubt that this mechanism generates meteorites; some meteorites in our collections must have originated on Mars-crossing asteroids — the unresolved problem is to determine which ones they are. Calculations indicate that this source can only supply a small fraction, perhaps a few percent, of the meteorites.

Figure IX-7 shows the frequency of asteroids between Mars and Jupiter. The great majority (and all those with radii greater than 30 km) have semimajor axes in the range 2.15 to 3.45 AU. Pronounced minima called Kirkwood gaps appear at certain semimajor axes. These gaps appear at distances where the orbital period corresponds to small-number fractions of the period of Jupiter. The fractions are listed along the top axis of the histogram; in the case of the fractions $\frac{2}{3}$ and $\frac{3}{4}$, peaks are found rather than gaps. The orbits of asteroids in period resonances with Jupiter undergo large changes in eccentricity (in the shape of the ellipse), but the aphelia never become large enough to permit close encounters with Jupiter that would lead to ejection from the solar system. Thus, these orbits are highly stable in a dynamic sense; however, when eccentricities are high, the Kirkwood-gap asteroids have high velocities relative to the other asteroids that occupy nearly circular orbits. The orbital stability is probably responsible for the peaks at semimajor axes greater than 3.6 AU, a region otherwise swept clean by nonresonant Jupiter interactions. The high velocities and consequent high kinetic energy release during collision with asteroids in circular orbits in the well-populated inner portion (less than 3.5 AU) of the belt has probably led to the preferential destruction of the asteroids missing from these Kirkwood gaps.

It is very difficult to transfer a meteoroid from the main belt into an Earth-crossing orbit as the result of an impact. In order to generate such a transfer, the velocity of the belt object must be changed by about 6 km · s^{-1}. This value is roughly equal to the mean velocity of impact between belt asteroids, but most of the material from an impact-produced crater is ejected with a velocity much less than the projectile's impact velocity. Moreover, most of the material ejected at high velocity is expected to be in the form of dust, and the minor amount of larger material should show more

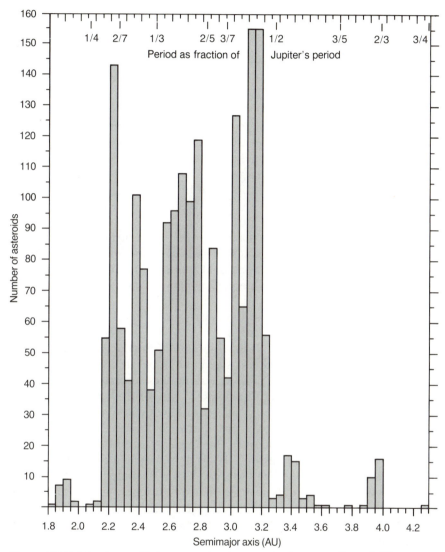

Figure IX-7. Most asteroids have semimajor axes in the range from 2.15 to
3.45 AU; this region (or a larger one extending from 1.8 to 4.3 AU) is known as
the asteroid belt. Closer to the Sun, asteroids have been removed by gravitational interactions with Mars, farther from the Sun by gravitational interactions
with Jupiter. Along the top of the diagram are marks indicating the positions of
semimajor axes yielding periods related to Jupiter's period as the ratio of small
integers. Inside 3.9 AU, minima called Kirkwood gaps appear near these
resonance locations, whereas peaks near 4.0 and 4.3 AU are associated with
these resonances.

severe shock effects than those typically observed in stony meteorites. If impacts were capable of producing efficient ejection of meteoroids having dimensions greater than 10 cm with velocities greater than the lunar escape velocity of 2.3 km · s^{-1}, it is remarkable that among 6000 meteorites we now have only two tiny (less than 35 g) ones from the Moon (see Chapter I). Thus single-impact transfer of asteroid-belt material to the Earth is not a viable mechanism.

A more-complex mechanism can transfer material into Earth-crossing orbits either from Kirkwood-gap asteroids or from another group of inner-belt (2.1 to 2.5 AU) asteroids having a precession-rate resonance coupled with the precession rate of the large planets. If, at times when the orbits of asteroids in either of these locations are highly elongated (that is, their eccentricities are high), cratered material is ejected with a velocity of more than 0.2 km · s^{-1} that removes it from the resonance, then further perturbations by Jupiter or Mars can make the ejected material Earth-crossing. The efficiency of this transfer mechanism is roughly defined; after allowance for the short (~ 20 Myr) attritional lifetimes, it appears that meter-sized debris knocked off these objects can only provide a few percent of the meteorites.

Another meteorite source that in many respects seems ideal is the Apollo asteroids. These bodies are already Earth-crossing (or nearly so), and thus their entire impact ejecta also is Earth-crossing. Furthermore, the escape velocities from these kilometer-sized objects are very low (only ~ 1.4R m · s^{-1}, where R is the radius in kilometers), and thus essentially all impact ejecta leave the object. The yield of meteorites from Apollo asteroids is uncertain, but the best current estimate is that most meteorites could be from this source. This model provides only a partial solution to the source problem, however. Because the Apollo-asteroid orbits are themselves unstable with halflives of ~ 10 Myr, a mechanism is required to replenish them from a long-lived source. Again, the only potential sources are comets and belt asteroids near resonance positions; no other source can generate the required rate, ~ 15 Apollos (radii greater than 0.5 km) each megayear. Mechanisms for extracting objects from resonant positions in the asteroid belt were discussed earlier.

Following a planetary perturbation, a new cometary orbit always passes through the spatial location where the perturbation occurred. Thus the aphelion of the new orbit generated by a perturbation by Jupiter will always be greater than 5.1 AU, Jupiter's orbital radius. Because Apollo asteroids have aphelia no greater than 4.2 AU, those that originated as comets cannot have had their current orbital parameters determined by a Jupiter interaction. One possibility is that a close interaction with a terrestrial planet has occurred, but such an interaction is expected only about once per megayear, and a close encounter with Jupiter will normally lead to ejection from the solar system before perturbation by a terrestrial planet can occur. The more-important mechanism appears to be a jet-type force resulting from evaporation from a spinning nucleus. Just as a spot on the Earth's surface is warmer at 1 or 2 P.M. than at noon, a comet is hottest (and evaporating most rapidly) not at the subsolar point (the point where the comet's surface intersects a line from the center of the comet to the Sun) but at a location generated by the rotation of that point. The extra force increases the aphelion distance and lengthens the period of the comet if the nucleus rotates in the same direction as the comet moves about the Sun, and the period is shortened (and the aphelion distance and semimajor axis decreased) if the nuclear rotation is opposite to the direction of orbital motion. The latter kind of interaction seems to be that required to produce Apollo-asteroid-type orbits from Jupiter-crossing comets. Because the rotational axes of comets are randomly distributed, one-half of them will have their semimajor axes decreased by jet action.

The ices in the exterior of a comet are almost entirely evaporated after $\sim 10^3$ passes through the inner solar system. Comet Encke now shows little activity even at perihelion, when it comes within 0.34 AU of the Sun; its orbital period is 3.3 yr. Evidence from the Taurid meteor streams has been used to infer that Encke has been in the inner solar system for about 5000 yr. After another 100 to 200 yr, the amount of gas emitted will be so small that Encke will be indistinguishable from an asteroid. The parent comet of the Geminid meteor shower was recently discovered by IRAS, the Infrared Astronomy Satellite (see Table IX-4). It has no coma and thus is technically no longer a comet.

Current estimates are that about 10 percent of the Apollo objects are from the asteroid belt, the remainder from comets. Uncertainties in both estimates are high, and it could be that as many as 50 percent originate in the asteroid belt. Although 50 to 90 percent of the Apollo asteroids may have originated as comets, we do not know whether any meteorites originated from these cometary asteroids; although some comets (such as the large Taurid meteoroids) contain silicate blocks, there is as yet no evidence indicating whether any cometary meteoroids had a combination of atmospheric velocity and strength such that they could survive atmospheric passage.

In summary, the favored conclusion is that most meteorites are fragments of the Apollo asteroids that were derived from belt asteroids, but that some are certainly from Mars-crossing asteroids, and that a small fraction may originate in comets. More important for understanding the relationship with the planets, however, is an understanding of where these parental asteroids and comets originated.

There seems to be no doubt that the comets originated beyond the orbit of Jupiter, because only in this region were nebular temperatures low enough to permit the condensation of H_2O. At any one location, the properties of the comet (such as the ratio of ice to rock) may have varied with the time of formation, and the detailed chemical composition of the ice probably varied with the distance from the Sun because, the greater the solar distance, the greater the number of volatile compounds that could have condensed before agglomeration of planetesimals occurred. The best working hypothesis appears to be that most comets in the Oort cloud formed in the region between ~ 15 and ~ 35 AU, were perturbed by interactions with Uranus and Neptune into orbits having aphelia of $\sim 10^5$ AU, and then (before another close encounter with these planets) interactions with passing stars increased the perihelia of some of these orbits converting them into Oort-cloud orbits.

Most of the cometary objects that formed in the region between 5 and 15 AU were ejected from the solar system by Jupiter and Saturn. At any moment in time, the fraction of these objects in orbits having aphelia of $\sim 10^5$ AU was very small, and thus few were transferred by stellar perturbations into the Oort cloud.

As discussed in Chapter VIII, it is doubtful that the initial formation of planetesimals by gravitational collapse of solid particles could have produced tough silicate rocks capable of surviving atmospheric passage. Adequately strong materials probably were produced mainly by impact-generated compression and heating near the surfaces of parent bodies having radii greater than 50 km, and to a lesser degree by hydrostatic compression in the hot interiors of parent bodies. Such processes certainly occurred in the inner solar system, but we have yet to learn whether they also occurred in the regions where comets formed.

The formation processes that produced the inner planets probably evolved as follows. Planetesimals having dimensions of 100 to 1000 m formed by gravitational collapse and gradually grew by low-velocity collisions. A few protoplanets grew large enough ($R > 100$ km) to have gravitational fields that could "stir" the orbits of the smaller objects — that is, could increase their mean eccentricities, and thus increase the width of the "feeding zone" from which the largest bodies could accumulate smaller bodies. Eventually, the inner solar system consisted of the embryos of today's four inner planets and a huge swarm of unaccreted planetesimals having radii in the range from 100 to 1000 km. Some of these planetesimals underwent repeated encounters with the protoplanets, resulting in the diffusion of planetesimals throughout the inner solar system. Although each planet largely formed from materials formed locally, each also accumulated a significant fraction of "foreign" planetesimals originally formed in the chief feeding zones of other planets.

Perturbations by the two large terrestrial planets Venus and Earth also brought planetesimal formed inside 1 AU (and probably as near the Sun as Mercury, $a = 0.38$ AU) into the asteroid belt. An additional perturbation was required in order to place these into long-term "storage" orbits. This perturbation was produced either by a close encounter with Mars that left the planetesimal in a Mars-crossing orbit or by a collision with an asteroid of similar or greater size. In most cases, a collision would have shattered both objects, but some residual fragments may still have had dimensions of 10 to 100 km.

An important feature of the use of a Mars encounter to decouple planetesimals from Earth and Venus is that this leaves such planetesimals in Mars-crossing orbits or (with an assist from Jupiter and another minor collision) in orbits having semimajor axes between 2.1 and 2.5 AU that are in precession resonances. Although an appreciable fraction of the bodies originally formed near Mars would experience similar dynamic histories, asteroids that formed inside ~1.8 AU may provide a sizable fraction of the subset of asteroids that, as discussed earlier, could be the dominant source of the Earth-crossing asteroids that are parental to meteoroids.

Although there does not appear to be any mechanism having appreciable efficiency that can inject entire outer-solar-system objects into the asteroid belt, collisions between trans-Jupiter objects and belt asteroids will leave some chunks of the trans-Jupiter materials in asteroid regoliths or in nearly circular orbits from which they could accrete onto asteroid surfaces.

Let me summarize this section. The largest source of meteorites is probably Apollo asteroids, although small fractions probably come directly from Mars-crossing asteroids, belt asteroids, and (if strong enough) comets. The Apollo asteroids have short dynamic halflives and are probably derived from objects in long-term storage orbits. Although most large Earth-crossing objects may originate as comets, the long-term (more than 10^9 yr) storage of the subset of Apollos parental to most meteoroids strong enough to produce meteorites is probably in or near asteroid orbits having resonances with Jupiter or in orbits that can be significantly perturbed by Mars. A sizable fraction of the asteroids in Mars-crossing orbits or in resonance locations between 2.1 and 2.5 AU may have initially formed inside 1.8 AU.

Proposed Formation Locations of Meteorites

Most of the following discussion is focused on chondritic meteorites because these best preserve the compositional clues to their formation locations. Of course, those clues that have survived in the differentiated meteorites will be used as much as is practical.

As shown in Table II-1, there is a wide range of properties between the extremes of the chondrites. It seems probable that this range encompasses most of that originally present in the solar system. This argument rests on the large compositional variations observed, which are difficult to explain in a small portion of the nebula (say, in that now corresponding to the asteroid belt, 2.15 to 3.45 AU from the Sun), and on dynamical arguments and calculations given earlier, indicating that materials formed anywhere between Mercury and Neptune could supply meteorites to the Earth.

As discussed in more detail in Chapter VIII, it is probable that both pressure and temperature increased sunward in the solar nebula. Thus, at any moment, the phases condensing nearer the Sun were more refractory than those condensing farther from the Sun. For this reason, at any time the equilibrium $FeO/(FeO + MgO)$ ratio in the mafic minerals increased with increasing distance from the Sun. Because of the effects of pressure, the temperature at which each phase condensed decreased with increasing distance from the Sun, and because diffusion rates increase with temperature, this probably led to a decrease in mean grain size with increasing solar distance (exceptions to this generalization could result if there were random variations in the degree of supercooling before nucleation at different locations).

Dynamical calculations show that the time at which particles in the median plane become unstable with respect to gravitational collapse increased with increasing distance from the Sun. Thus, the planetesimals that formed nearest the Sun are expected to have had the lowest $FeO/(FeO + MgO)$ ratio and the largest mean grain size. Depending on the temperature at the time the planetesimals formed, certain elements were essentially uncondensed and others only partially condensed.

On the basis of such arguments, most cosmochemists have concluded that, in chondrites containing metallic $Fe-Ni$, increasing $FeO/(FeO + MgO)$ ratio in chondrites should be a moderately reliable indicator of relative distance from the Sun. As shown in Figure IX-8, refractory-element abundances correlate roughly with $FeO_x/(FeO_x + MgO)$. The refractory-abundance trend may reflect an increasing tendency of the refractory portion of interstellar materials to survive evaporation with increasing distance from the Sun, but the cosmochemical arguments for such a distance dependence are not as compelling as are those involving the $FeO_x/(FeO_x + MgO)$ ratio.

If the general sequence from lower left to upper right in Figure IX-8 is one of increasing distance from the Sun, we need to define the distance scale. The Earth is the planet we know best, and representing the Earth are two circles. In Ringwood's pyrolite (a hypothetical upper-mantle material), the refractory abundance is influenced by his belief that the Earth has CI-chondrite abundances of most nonvolatile elements. Some relatively primitive upper-mantle rocks have calcium and aluminum abundances lower than that in pyrolite, and $Rb-Sr$ and $Sm-Nd$ systematics in recent volcanic rocks are more-easily understood if refractory abundances in the whole Earth are more like those in ordinary chondrites. The circle drawn as

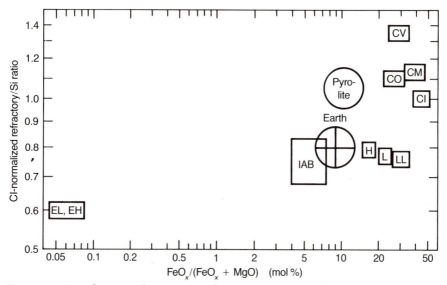

Figure IX-8. Refractory-element abundances in the chondrite groups correlate roughly with degree of oxidation as indicated by $FeO_x/(FeO_x + MgO)$ ratios; the latter ratio is expected to increase with increasing distance from the Sun. Two circles show estimates for the bulk Earth composition: Ringwood's pyrolite, and a composition having the lower refractory abundance listed in Table IX-2. Thus, this diagram suggests that the enstatite clan formed nearest the Sun and that the IAB chondrites are the group formed nearest 1 AU.

the Earth symbol shows the ordinary-chondrite-like refractory abundance and the slightly lower $FeO_x/(FeO_x + MgO)$ ratio listed in Table IX-2. The difference in terrestrial refractory abundances between the two circles is indicative of the uncertainty of ~ 20 percent associated with current estimates.

The position of the Earth on Figure IX-8 suggests that the chondrites that formed nearest the Earth are the H and/or IAB chondrites. The latter may be the more closely related, because correction of the mantle $FeO_x/(FeO_x + MgO)$ ratio downward to allow for late-accreted chondritic material results in an Earth ratio only slightly higher than those in the IAB mafic silicates.

In Figure IX-9, the oxygen-isotope compositions of the chondrites, the Earth and the Moon are plotted; note that this is an expanded portion of Figure III-9. Recall that chemical or physical fractionation produces trends having a slope of 0.52. The oxygen-isotope compositions of all terrestrial samples lie along a line segment having a slope of 0.52 that extends past the edges of this diagram.

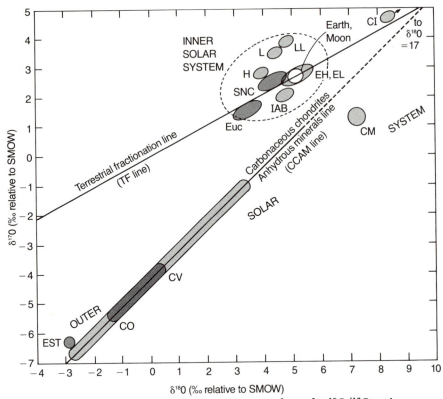

Figure IX-9. Fractionation of the oxygen isotopes alters the $^{18}O/^{16}O$ ratio (represented by $\delta^{18}O$) twice as much as the $^{17}O/^{16}O$ ratio (represented by $\delta^{17}O$) and produces an array having a slope of 0.5. The line labeled "terrestrial fractionation" results from fractionation processes in terrestrial systems. The field marked "Earth, Moon" shows the estimated compositions of the mantles of these bodies. Chondrites having compositions similar to the Earth are inferred to have originated in the inner solar system. Materials plot away from the terrestrial line if they contain appreciable amounts of unevaporated interstellar solids having anomalous oxygen isotope compositions. Because the fraction of interstellar matter that escaped evaporation increases with distance from the Sun, the CV, CO, and CM groups are inferred to have formed in the outer solar system. See text for a discussion of the CI group. (Data from publications by R. N. Clayton, T. Mayeda, and colleagues.)

The anhydrous minerals (olivine, pyroxene, and refractory minerals) in CM, CO, and CV chondrites form a linear array with a much-higher slope of about unity that extends from $\delta^{18}O = -42\%_0$ to about $+5\%_0$. Such an array cannot be produced by fractionation processes (see Chapter III for discussion of an exception) but requires the mixing of material from two isotopically distinct reservoirs.

If the temperature in the solar nebula fell with increasing distance from the Sun, one would expect that near the Sun a greater fraction of the precursor interstellar solids were vaporized than at more distant locations. Thus the preservation of large oxygen-isotopic anomalies in CM, CO, and CV chondrites implies that they formed relatively far from the Sun.

Because the Earth and Moon formed mainly from materials that originated near 1 AU in the inner solar system, their position on Figure IX-9 should mark the general location of inner-solar-system solid materials. The chondrites having low $FeO_x/(FeO_x + MgO)$ ratios and low refractory abundances tend to cluster about the Earth–Moon field, implying that they also formed in the inner solar system. Thus the chemical (Figure IX-8) and the oxygen-isotopic evidence independently support the division of the chondrites into two broad categories: inner-solar-system materials (enstatite, IAB, and ordinary clans) and outer-solar-system materials (refractory-rich, minichondrule, and volatile-rich clans). The inclusion of the volatile-rich (CI) chondrites among the outer-solar-system materials is based primarily on the chemical evidence. The oxygen-isotope data plot near the terrestrial-fractionation line but, if one recalls that these objects consist entirely of fine-grained minerals, the arguments given in the footnote[1] show that the oxygen-isotope data are not inconsistent with their formation in the outer solar system.

There are additional clues regarding both the absolute and the relative

----

[1] A large fraction of the oxygen in the solar nebula and in the precursor interstellar cloud was in the form of gaseous H_2O. In the nebula, about one-half of the oxygen was present as H_2O; in the cloud, the fraction was probably higher, because much of the carbon formed CH_4, and thus less of the oxygen was bound to carbon (as CO). Because temperatures were high in the inner solar system, it seems certain that in this region the oxygen-isotopic composition of H_2O, CO, and any other important reservoirs were near the terrestrial-fractionation line, and quite probable that the mean interstellar H_2O isotopic composition was similar. Only if a large oxygen reservoir other than H_2O (CO seems to be the only candidate) had an oxygen-isotopic composition well away from the terrestrial-fractionation line would interstellar H_2O have deviated in the opposite direction. The most likely explanations of the oxygen-isotopic composition of the fine-grained, hydrous silicates in the CI chondrites are (1) that they were determined by formation from, or exchange of silicate oxygen with, gaseous H_2O having a composition near the terrestrial line; or (2) that they reflect the mean compositions of the fine interstellar dust that failed to evaporate at that distance from the Sun. Because there is evidence in the ordinary chondrite chondrules and CV, CO, and CM refractory minerals for very different oxygen-isotopic compositions in the presolar solids, it seems more likely that the composition of CI chondrites resulted from interaction with nebular gases.

distances from the Sun. The details of some of the evidence summarized next were discussed earlier in this chapter.

One model to account for the elemental fractionation pattern of the rare gases in the Venus atmosphere calls for solar-wind irradiation of nebular matter near the Sun at a time when nebular gases had dissipated in this region but were still present at distances greater than 0.8 AU from the Sun. The similarity in rare-gas fractionation patterns between Venus and the enstatite-clan chondrites thus supports the hypothesis that these planetesimals formed inside the orbit of Venus.

If the enstatite chondrites formed within 0.7 AU of the Sun, then the lavas on the surface of Mercury should have very low $FeO/(FeO + MgO)$ ratios.[2] As discussed earlier, remote spectral observations of the Mercurian regolith indicate a stronger pyroxene FeO absorption than that expected from an enstatite-chondrite model. Poor resolution, however, makes evaluation of the FeO content difficult; also, some FeO in the Mercurian regolith must have resulted from the late accretion of highly oxidized chondritic matter.

As discussed earlier, the regolith $FeO_x/(FeO_x + MgO)$ on Mars implies a higher bulk-planet ratio than estimated for the Earth. However, the uncertainty in the bulk-Mars ratio (see Table IX-2) is too large to permit the inference of genetic links between Mars and specific groups of meteorites. Nor does the lower limit of 8 on the FeO/MnO ratio in Mars offer any help in establishing links to chondritic groups. Furthermore, the genetic significance of this ratio remains uncertain until questions about the effect of manganese volatility on planetary manganese contents have been answered.

The division of the chondrites into inner- and outer-solar-system materials on the basis of oxygen isotopes was discussed earlier. It would be useful if more-exact locational information could be inferred from these data, but to date only rough inferences are possible. It seems plausible that the fine-grained materials of each chondrite group offer the best evidence regarding the nebula temperature at which agglomeration (planetesimal formation) occurred. Calculations based on laboratory measurements of the partitioning of isotopes between coexisting phases show that solids in equilibrium with the gas will have their $\delta^{18}O$ values displaced negatively at high temperatures and positively at low temperatures. A detailed discussion is not appropriate to this text but, if the plausible assumption is made

---

[2] A popular competing hypothesis is that the silicates of Mercury mainly consist of refractory materials similar to the Ca-Al-rich inclusions in CV chondrites. This model also predicts a low $FeO/(FeO + MgO)$ ratio but in addition a very-low abundance of alkalis on Mercury, whereas an enstatite chondrite model for Mercury yields a Na/Si ratio near that in CI chondrites.

that the mean $\delta^{18}O$ value of the bulk nebula was about $+8‰$, the data are consistent with essentially the same formational-distance sequence inferred from $FeO_x/(FeO_x + MgO)$ values: enstatite $<$ IAB $\cong$ Earth $<$ ordinary $\ll$ CM $=$ CO $=$ CV $<$ CI.

An important question is why some inner-solar-system groups plot on one side of the terrestrial-fractionation line and some on the other. As discussed in Chapter III, it appears that this division requires (1) the existence of isotopically different reservoirs of presolar solids, and (2) nebular mechanisms that allow differing ratios in the relative amounts of these solids processed and agglomerated under conditions that varied with time and place.

As documented in Appendix A, the seven groups of meteorites that fall most frequently are L chondrites (39 percent of falls), H chondrites (32 percent), LL chondrites (7.2 percent), eucrites (2.8 percent), howardites (2.8 percent), CM chondrites (2 percent), and IIIAB irons (1.5 percent). None of the remaining groups have fall frequencies significantly greater than 1 percent but, because of their extreme friability,[3] the CI chondrites (0.7 percent) deserve inclusion in this discussion. Although it is not possible to apply even a semiquantitative bias correction, a relative ordering in terms of decreasing crushing strength would be: IIIAB irons $\gg$ ordinary chondrites $>$ eucrites $=$ howardites $>$ CM chondrites $>$ CI chondrites. Thus, bias corrections would draw the fall frequencies of the seven groups of stony meteorites closer together and probably would show CI chondrites to fall significantly more frequently than IIIAB irons.

What is the significance of the fact that the three closely related groups of ordinary chondrites dominate the fall statistics? This clearly means that these three (or possibly more) parent bodies, after forming at a narrow range of distances from the Sun, are now in storage orbits that are situated such that they can produce meteoroids with a high relative efficiency. As discussed earlier, many or most asteroids in the center of the belt probably formed in essentially the same orbits they now occupy, whereas the Earth-crossing and many Mars-crossing asteroids, as well as some having orbits linked to Jupiter by resonances, could have formed at other locations quite distant from the present locations. In the latter case, objects now in similar orbits could have diverse compositions. The fact that the ordinary chondrites are both closely related and the three most common groups to fall

---

[3] CI chondrites are so friable that they can be crushed between the thumb and forefinger, and it is likely that the fraction of these materials surviving atmospheric passage is low. An example of their weakness is the fact that microbarometric observations of the fall of the Revelstoke CI chondrite indicated a mass of $\sim 10^8$ g $= 10^2$ t (assuming an atmospheric velocity of 20 km $\cdot$ s^{-1}), whereas despite extensive searches only 1 g was recovered.

seems most easily understood if they originally formed in the main portion of the asteroid belt 2.2 to 3.4 AU from the Sun. Perhaps each parent body was associated with a different Kirkwood gap.

The balance of the evidence (Chapter V) suggests that the igneous-clan differentiated meteorites (eucrites, howardites, and so on) formed relatively near the Sun. The increase in nebular temperature and the general increase in rate of agglomeration and accretion led to an increase in the probability of melting with decreasing solar distance. Thus, a plausible scenario is that the parent body of the igneous clan formed near (less than 1.8 AU from) the Sun and was stored in a Mars-crossing or inner-belt resonance orbit prior to its destruction to form the igneous-clan debris.

Because of their highly unequilibrated, hydrated silicates, their high degree of oxidation, and their high contents of volatiles, the CM and CI chondrites are the strongest candidates to be cometary debris. If they are from comets, the six (probably seven, including CI chondrites) most-common groups of meteorites include representatives of the three categories of formation locations cis-Mars (less than 1.5 AU), asteroid belt, and trans-Jupiter (greater than 5 AU), and the source strengths of these three kinds of meteoroids are the same to within a factor of ~ 10. I stress again, however, that there is as yet no firm link of the CM or CI chondrites with cometary parent bodies.

There are two distinct major populations of asteroids: the low-albedo C objects, and the high-albedo S objects. It is highly improbable that the observed surface properties of these objects could have formed by a simple nebular-fractionation mechanism that varied systematically and continuously with time or solar distance. If the different surface properties reflect differences in the bulk materials, then the two most plausible explanations of the observations are (1) that one population formed in the asteroid belt, and the other formed elsewhere; or (2) that one population is undifferentiated (chondritic), and the other formed from these materials by melting and differentiation. The second model seems unlikely because it is doubtful that there were suitable heat sources for differentiating asteroids at orbits larger than 3 AU, and because the expected tendency for the largest asteroids to be differentiated is not observed. The first model leads to the expectation that "foreign" asteroids should be concentrated near resonances or in Mars-crossing orbits. Although there is a tendency for albedo to decrease toward the Sun (consistent with a model calling for S asteroids to form less than ~ 1.8 AU from the Sun, searches have not shown a tendency for S asteroids to have semimajor axes near resonance values.

Another possibility is that the surface properties of an appreciable fraction of the asteroids are not representative of the bulk compositions. In this case, the following model becomes plausible: most of the mass in the asteroid belt formed by agglomeration and accretion early (during the first 0.1 to

0.3 Gyr) in solar-system history but, during a much longer period (until ~3.7 Gyr ago), appreciable amounts of outer-solar-system material were scattered by the gravitational fields of the major planets through the entire solar system (a small part went into the Oort cloud of comets), and the collisional disruption of this material in the asteroid belt produced large amounts of dust and chunks that accreted onto the surfaces of preexisting asteroids as a deep layer several kilometers deep. The early material could be ordinary chondrites; the late-accreted matter in pure form might resemble CM chondrites — in regoliths, it might be a component in the solar-noble-gas-rich dark portions of the light – dark breccias (see Chapter III). According to this model, the C asteroids should have S-type "cores," so we might expect asteroids that are irregular in shape (indicating fragmentation), that are members of genetically related families produced by fragmentation, or that are small (suggesting either fragmentation or a low escape velocity and inefficient ability to accrete the dust) to tend to be S-type. There is evidence for a higher abundance of S asteroids in each of these categories, but the S fractions are not so dominant that one can rule out other possible explanations.

Because of the close similarity in reflection spectra, a popular hypothesis is that the igneous eucrites and howardites originated on Vesta. There are strong arguments against this, however. (1) All evidence seems consistent with the interpretation that the eucrites and howardites originated in the same parent body with the pallasites (and possibly the IIIAB irons); because these latter materials require excavation to the core – mantle interface, they cannot originate in Vesta, whose surface is still intact. It is probable that the parent body of the igneous clan was disrupted and now exists only as fragments. (2) The eucrites and howardites are among the most-common meteorites classes to fall. Given similar orbital properties, the strength of an asteroidal source of meteoroids should be proportional to its surface area and inversely related (in complex fashion) to its escape velocity. Because the many small fragments of a broken igneously fractionated asteroid should have a much-higher cumulative surface area and far-lower escape velocities, such objects (if they exist) could provide many times more meteoroids than Vesta could.

Many solar-rare-gas-rich breccias include clasts of foreign chondritic material. In most cases, detailed investigation of these materials has shown them to be closely related to CM chondrites in texture and composition; the clasts generally have lower contents of some volatiles, especially H_2O. Thus, when these breccias formed more than 3.6 Gyr ago, CM chondrites were the dominant type of foreign interplanetary matter at the inner-solar-system locations where these regoliths were evolving. The siderophiles in the lunar regolith preserve a cumulative record of the composition of interplanetary materials falling since the regolith started to form. In the maria, the youngest (less than 3.8 Gyr) portions of the lunar surface, siderophile ratios are more similar to those in CM chondrites than to those in any other meteorite group, implying that CM-chondrite-like matter has been the most common kind of material[4] in Earth-crossing orbits during the past 3.8 Gyr. As noted above, at 2 percent of observed falls, CM chondrites are the sixth-most-common group. Allowance for their low crushing strength indicates that a substantially higher fraction strikes the upper atmosphere. Furthermore, if the CM-like materials are in "cometary" orbits, then their mean atmospheric velocities are probably much higher than those of the ordinary chondrites. Thus it seems quite possible that their abundance (and that of CI-chondrite-like materials as well) is comparable to or greater than that of the ordinary chondrites. Of course, matter striking the atmosphere-less Moon is totally destroyed, and thus we cannot distinguish between weak meteoroids and those tough enough to reach the Earth's surface as meteorites, so the large fraction of CM-like materials contributing siderophiles to the lunar regolith provides only an upper limit to the fraction of meteoroids large enough and strong enough to produce meteorites.

None of the evidence regarding meteorite formation locations is conclusive, and much is only suggestive. Because of the importance of the topic it is nonetheless important to keep the available evidence in mind and to develop working hypotheses that can be tested as additional data become available. It appears that the best working hypothesis has the following components.

1. The chondrites can be divided into inner- and outer-solar-system materials.
2. The inner-solar-system chondrites formed at locations increasing in distance from the Sun through the clan sequence enstatite – IAB – ordinary.
3. There is little evidence that can be used to infer the sequence among outer-solar-system chondrites, but a best estimate of increasing mean distance is CV < CO < CM < CI; some (especially CM and CO) may have formed sequentially at the same location.
4. The division between inner- and outer-solar-system formation locations is either in the asteroid belt or near the orbit of Jupiter. The weight of the evidence favors the latter, in which case it is probable that the ordinary chondrites formed in the asteroid belt.

---

[4] A caveat is necessary here; because only a few siderophiles are well determined in the lunar regolith, a mixture of CI chondrites with roughly equal parts of a chondritic material having low volatile contents (such as CV chondrites) would be difficult to distinguish from CM chondrites.

How can evidence be obtained in the future to determine just where the meteorites formed? The best-possible approach would be to use a spacecraft to obtain samples from asteroids or comets and return them to Earth for exact characterization. The best target asteroid would be undifferentiated, irregular in shape (thus likely to have materials from the earliest agglomeration periods exposed on its surface), and in a highly stable circular orbit (thus likely to have formed at its present location). The best comet would be one like comet Encke that is known to contain large, relatively strong silicate masses that, given a low atmospheric velocity, might be capable of falling as meteorites; a comet such as Giacobini–Zinner in which most solid meteoroids seem to be extremely weak may not offer materials well suited for comparison with meteorites.

Spacecraft missions capable of sample return are expensive, and thus it may be several decades before such missions take place. The next-best thing is a landing spacecraft that gathers data *in situ*, but such a mission will not generate some of the most important data—for example, data on oxygen isotope composition, other isotopic systems, rare trace elements, or detailed petrology.

A much less expensive alternative to a spacecraft mission would be an improvement in the recovery of meteorites having well-determined orbits. If inexpensive cameras with electronic-digital recording systems could be developed, these could be placed in arrays that are more closely spaced than the ~50-km separations in the current networks. Digital recording could eliminate mechanical parts and allow rapid trajectory determinations. Closer spacing could increase the chance of recovery and make it worthwhile to search for meteorites having end masses as small as 100 g, in contrast to the ~1-kg limitation of the existing networks. Particularly rewarding would be the recovery of a meteorite having an aphelion greater than 5.2 AU, because such a meteorite would almost certainly have originated in a comet.

Suggested Reading

Chapman, C. R. 1975. The nature of asteroids. *Sci. Amer.* **232**(1):24–33. A nontechnical discussion of asteroid observations with emphasis on reflection spectra, and a discussion of the formation and collisional evolution of the asteroid belt.

Chapman, C. R., J. G. Williams, and W. K. Hartmann. 1978. The asteroids. *Ann. Rev. Astron. Astrophys.* **16**:33–75. A technical review of asteroid observations, orbital properties, and formation models.

Ringwood, A. E. 1979. *Origin of the Earth and Moon.* Springer. A technical discussion of the measured properties of the terrestrial and lunar rocks and methods by which these can be inferred to yield the bulk properties of these planets.

Wasson, J. T., and G. W. Wetherill. 1979. Dynamical, chemical and isotopic evidence regarding the formation locations of asteroids and meteorites. In *Asteroids*, ed. T. Gehrels, pp. 926–974. University of Arizona Press. Technical discussion of the dynamical, chemical, and isotopic evidence regarding the formation locations of asteroids and large meteorites.

Wetherill, G. W. 1974. Solar system sources of meteorites and large meteoroids. *Ann. Rev. Earth Planet. Sci.* 2:303. A technical discussion of the orbital and, less extensively, the compositional evidence regarding the solar-system sources of meteorites and large meteoroids.

Wetherill, G. W. 1979. Apollo objects. *Sci. Amer.* **240**(3):54–65. A relatively nontechnical discussion of the Earth-crossing Apollo asteroids and an evaluation of belt asteroids and comets as sources.

Whipple, F. L. 1978. Comets. In *Cosmic Dust*, ed. J. A. McDonnell, pp. 1–72. Wiley. Moderately technical discussion of observations of comets and inferences regarding cometary properties.

APPENDIX A

Meteorite Classification

Table A-1 lists every meteorite group that contains five or more members and shows the classification into clans that is used in this book.

Table A-1
Classification of meteorites into clans and groups

Clan	Group	Synonyms[°]	Example	Falls	Finds	Fall frequency[†] (%)
Chondrites:						
Refractory-rich	CV chon. ⎱	C3 carbonaceous	⎰ Allende	8	3	1.1
Minichondrule	CO chon. ⎰		⎱ Ornans	6	1	0.85
	CM chon.	C2 carbonaceous	Murchison	14	0	2.0
Volatile-rich	CI chon.	C1 carbonaceous	Orgueil	5	0	0.71
Ordinary	LL chon.	Amphoterite chon.	St. Mesmin	51	16	7.2
	L chon.	Hypersthene chon.	Bruderheim	278	192	39.3
	H chon.	Bronzite chon.	Ochansk	229	230	32.3
IAB-inclusion	IAB chon.	—	Copiapo	—[‡]	—[‡]	—[‡]
Enstatite[§]	EL chon.	Enstatite	Indarch	6	3	0.85
	EH chon.	Enstatite	Khairpur	5	3	0.71
—	Other chon.	—	Kakangari	2	2	0.28
Differentiated meteorites:						
Igneous	Eucrites (EUC)	Basaltic, calcium-rich, or pyroxene–plag. achon.	Juvinas	20	4	2.8
	Howardites (HOW)	Same as EUC	Kapoeta	18	1	2.5
	Diogenites (DIO)	Hypersthene achon.	Johnstown	8	0	1.1
	Mesosiderites (MES)	Stony-irons	Estherville	6	14	0.85
	Pallasites (PAL)	Stony-irons	Krasnojarsk	2	33	0.28
—	Ureilites (URE)	Olivine–pigeonite achon.	Novo Urei	3	3	0.42
Enstatite[§]	Aubrites (AUB)	Enstatite achon.	Norton County	8	1	1.1

		7	6	0.99
Other differentiated silicate-rich meteorites	Shergotty	—	—	—
IAB irons	Canyon Diablo	7	83	0.85
IC irons	Bendegó	0	10	0.09
IIAB irons	Coahuila	5	47	0.49
IIC irons	Ballinoo	0	7	0.06
IID irons	Needles	2	11	0.12
IIE irons	Weekeroo Station	1	11	0.11
IIF irons	Monahans	0	5	0.05
IIIAB irons	Henbury	6	151	1.46
IIICD irons	Tazewell	2	10	0.11
IIIE irons	Rhine Villa	0	8	0.08
IIIF irons	Nelson County	0	5	0.05
IVA irons	Gibeon	2	38	0.38
IVB irons	Hoba	0	11	0.10
Other irons	Mbosi	6	62	0.62

Abbreviations: chon. = chondrites; achon. = achondrites; plag. = plagioclase

° Although not used in this book, some of these synonyms are quite common in the current literature. No relevant synonyms exist for the groups of irons.

† Fall frequencies of iron meteorites are calculated by arbitrarily allocating the 32 observed falls confirmed by V. F. Buchwald (*Iron Meteorites,* University of California Press, 1975) to the frequencies of all irons classified by Scott and Wasson (*Rev. Geophys. Space Phys.* 73:530, 1975).

‡ See data for IAB irons.

§ The enstatite-clan chondrites and the aubrites are very closely related.

‖ Clan relationships involving irons are still poorly understood. The only known close relationships are between IIIAB and IIE irons (possibly also including the pallasites) and between the IIE irons and the H chondrites. Although the structural classification of irons into hexahedrites, octahedrites, and ataxites does not lead to genetically related groups, *most* of the fine octahedrites belong to group IVA, *most* of the medium octahedrites to group IIIAB, *most* of the coarse octahedrites to subgroup IA, and *all* of the large-crystal (at least 5 cm) hexahedrites to subgroup IIA.

APPENDIX B

Units and Constants

Table B-1 lists the units used in this book; it also includes the value of the gravitational constant G.

The International System of Units (Système International d'Unités, or SI) was established in 1960 by international agreement. Seven SI base units are defined in terms of actual physical phenomena: the meter (metre), the kilogram, the second, the ampere, the kelvin, the mole, and the candela (a unit of luminous intensity). All other SI units are derived units that are defined terms of the base units. Table B-2 lists the prefixes that can be used with any SI unit to indicate multiples or submultiples of that unit.

Some non-SI units are still used commonly in astronomical literature; the SI equivalents of such units used in this book are given in Table B-1.

Table B-1
Units and constants

Unit or constant	Abbreviation	Quantity	Equivalent to
Atmosphere	atm	Pressure	$10^5 \, \text{Pa} = 10^5 \, \text{N} \cdot \text{m}^{-2}$
Astronomical unit	AU	Length	1.50×10^{13} cm
Electron volt	eV	Energy	1.60×10^{-19} J
Gram	g	Mass	10^{-3} kg
Gravitational constant	G	——	$6.67 \times 10^{-8} \, \text{cm}^3 \cdot \text{g}^{-1} \cdot \text{s}^{-2}$
Joule	J	Energy	$1 \, \text{m}^2 \cdot \text{kg} \cdot \text{s}^{-2} = 1 \, \text{N} \cdot \text{m}$
Kelvin	K	Thermodynamic temperature	
Mass of Earth	$M_\oplus$	Mass	5.98×10^{27} g
Mass of Sun	$M_\odot$	Mass	1.00×10^{33} g
Meter (metre)	m	Length	100 cm
Mole	mol	Amount of substance	
Newton	N	Force	$1 \, \text{m} \cdot \text{kg} \cdot \text{s}^{-2}$
Parsec	pc	Length	3.09×10^{18} cm
Tonne (metric ton)	t	Mass	10^6 g
Second	s	Time	——
Year	yr	Time	3.16×10^7 s

Table B-2
SI prefixes

Prefix	Abbreviation	Multiplier
pico-	p	10^{-12}
nano-	n	10^{-9}
micro-	μ	10^{-6}
milli-	m	10^{-3}
centi-°	c	10^{-2}
kilo-	k	10^{3}
mega-	M	10^{6}
giga-	G	10^{9}
tera-	T	10^{12}

° Centi-, while not an SI prefix, is used in the text and is listed for convenience.

APPENDIX C

Common Meteoritic Minerals

Table C-1 lists the minerals most commonly found in meteorites and gives their chemical formulas.

Table C-1
The most-common meteoritic minerals

Mineral	Formula	Remarks
Albite	$NaAlSi_3O_8$	Component of plagioclase
Anorthite	$CaAl_2Si_2O_8$	Component of plagioclase
High-calcium clinopyroxene	$Ca_xAl_{2y}(Mg,Fe)_{1-x-y}Si_{1-y}O_3$	$x \cong 0.45; y \leq 0.05$
Low-calcium clinopyroxene	$Ca_x(Mg,Fe)_{1-x}SiO_3$	$x \leq 0.05$ (see pigeonite)
Cohenite	$(Fe_xNi_{1-x})_3C$	$x \cong 0.9$
Diopside	$Ca_{0.5}Mg_{0.5}SiO_3$	A form of high-calcium clinopyroxene
Fayalite	Fe_2SiO_4	Component of olivine
Ferrosilite	$FeSiO_3$	Component of pyroxene
Forsterite	Mg_2SiO_4	Component of olivine
Kamacite	Fe_x-Ni_{1-x}	$0.96 \leq x \leq 0.93$
Oldhamite	CaS	——
Olivine	$(Mg,Fe)_2SiO_4$	See fayalite, forsterite
Orthopyroxene	$Ca_x(Mg,Fe)_{1-x}SiO_3$	$x \leq 0.05$
Pigeonite	$Ca_x(Mg,Fe)_{1-x}SiO_3$	$x \cong 0.1$; a clinopyroxene
Plagioclase	$(NaSi)_x(CaAl)_{1-x}AlSi_2O_6$	$0.1 < x < 0.9$
Schreibersite	$(Fe_xNi_{1-x})_3P$	$x \cong 0.7$
Taenite	Fe_x-Ni_{1-x}	$0.5 \leq x \leq 0.8$
Troilite	FeS	——

APPENDIX D

Solar-Atmosphere and CI-Chondrite Atomic Abundances

The CI carbonaceous chondrites are the chondrites having the highest contents of volatiles. As shown in Figure II-1, the most-accurately determined solar abundances[1] of volatile elements show good agreement with volatile abundances in CI chondrites but are significantly higher than those in CM chondrites (the group having the second-highest abundances of volatiles).

The CI chondrites can be studied in the laboratory, and their elemental concentrations (with few exceptions) have been determined with good accuracy. Silicon-normalized CI abundances are accurate to ± 10 percent. Solar abundances are determined mainly by telescopic spectroscopic studies of lines produced by absorption or emission at various depths (and pressures and temperatures) in the hot solar atmosphere, and these values are rarely precise to better than ± 20 percent. For this reason, data on the CI chondrites are believed to offer the best estimates of mean solar or solar-nebula compositions except for the incompletely accreted, most-volatile elements: hydrogen, carbon, nitrogen, oxygen, and the rare-gas elements.

Table D-1 lists solar-abundance data from J. E. Ross and L. H. Aller and CI-chondrite data compiled especially for this text. The CI-chondrite data are based on a survey of the literature through June 1983. Because there is considerable variation in the quality of the data, a selection must be made. The references actually used are cited in Table D-1.

Another recent compilation by E. Anders and M. Ebihara[2] gives a more complete listing of relevant CI data sources and criteria for choosing values when data are lacking or show excessive scatter. Their selected CI concentrations for beryllium, boron, rhodium, tantalum, mercury, and the 14 rare-earth elements lanthanum through lutetium are used in Table D-1. My CI concentrations for the remaining 59 elements agree with those of Anders and Ebihara to within $+5$ percent with the exception of nine elements. Six of these (carbon, sulfur, yttrium, niobium, antimony, and osmium) agree to within ± 10 percent; for the remaining three, the ratios of their value to mine are 0.88 for nitrogen, 1.20 for phosphorus, and 0.87 for iodine.

---

[1] An abundance is the atomic ratio of the element of interest to a normalizing element, commonly silicon.

[2] E. Anders and M. Ebihara. 1982. Solar-system abundances of the elements. *Geochim. Cosmochim. Acta* 46:2363–2380.

One good method to test the quality of abundance data is to plot the logarithm of nuclide[3] abundance versus mass number and examine whether the tabulated abundances lead to smooth variations. Figure D-1 shows such a diagram for the mass-number range of 21 through 209. Below mass number 21, most elemental abundances have high uncertainties; between mass numbers 209 and 232, all nuclides have half-lives that are short compared to the 4.5 Gyr age of the solar system and, because of this large hiatus, points for the long-lived $(T_{1/2} > 1$ Gyr$)$ ^{232}Th, ^{235}U, and ^{238}U are not plotted.

Separate curves are described by the abundances of even-mass-number and old-mass-number nuclides (Figure D-1). The even-mass-number curve is generally higher by ~ 0.3 log units, but in a few cases the curves become superposed. The simplest explanation of this discrepancy is the following. At mass numbers of 36 or more, there are two or three stable nuclides for each even mass number whereas, with rare exceptions, there is only one stable nuclide for each odd mass number.

A more-correct explanation is that proton or neutron pairs are especially stable. Said another way, an "even" proton has about 2 MeV more binding energy[4] than does an "odd" proton, and the extra binding energy is similar for "even" neutrons. The stable even-mass-number nuclides thus have ~ 2 MeV more binding energy than do neighboring stable odd-mass-number nuclides, and this higher stability is an additional factor accounting for the higher integrated abundances of even-mass-number nuclides. An interesting byproduct of the difference in the binding energies of odd and even protons or neutrons is that odd-atomic-number, odd-neutron-number (odd–odd) nuclides are unstable with respect to beta decay to even–even nuclides; the few odd–odd nuclides above mass number 14 that exist in nature are radioactive.

Most heavy nuclides (mass number at least 60) were formed by neutron capture, and the abundance of a nuclide produced by this process is inversely proportional to its cross-section (a measure of the probability of neutron capture). The extra stability of the even–even nuclides results in

---

[3] As defined in Chapter III, a **nuclide** is an atomic or nuclear species characterized by any two of the three quantities atomic number, neutron number, and mass number. Thus $^{12}_{6}$C is a nuclide having 6 protons and 6 neutrons, which combine to give a mass number of 12.

[4] **Binding energy** is the difference in energy or mass between the free and the bound proton or neutron. For example, when a neutron is added by the reaction

$$^{A-1}_{Z}X + ^{1}_{0}n = ^{A}_{Z}X + \text{BE}$$

then BE is the binding energy of that neutron.

their average cross-sections being somewhat lower than those of neighboring odd-mass-number nuclides. In some regions the even-mass-number curve is zigzag, particularly at mass numbers 56 or less, where the capture of α particles (^{4}He nuclei) is an important mode of nucleosynthesis. The odd-mass-number curve is remarkably smooth; its peaks are round rather than jagged. In those cases where an element has two stable nuclides, the local slope is determined by their relative isotopic abundances, and these elements can sometimes be used to confirm general trends in particular mass-number regions. An example is the existence of the valley at mass number 135.

The various peaks appearing in Figure D-1 reflect high efficiencies of certain nucleosynthesis processes in these mass-number regions. For example, the peak near mass number 56 is attributed to an approach to "nuclear equilibrium" in the centers of very hot stars; these abundant nuclides have minimum values of mass/mass-number ratios, or maxima in binding energy per mass number. The peak near mass number 194 is attributed to the effect of a closed shell at neutron number 128 during neutron capture on a rapid time scale.

The scope of this book does not permit giving more details on this fascinating subject. An excellent treatise at an upperclass textbook level is that by J. Audouze and S. Vauclair.[5] A more-technical recent book is the volume of essays by experts in the field edited by C. A. Barnes et al.[6]

---

[5] Audouze, J., and S. Vauclair 1980. *An Introduction to Nuclear Astrophysics.* Reidel.

[6] Barnes, C. A., D. D. Clayton, and D. N. Schramm, eds. 1982. *Essays in Nuclear Astrophysics.* Cambridge University Press.

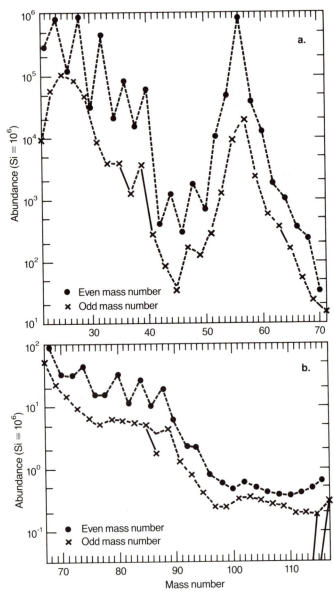

Figure D-1. Solar abundances of elements plotted against the mass number (the number of protons plus neutrons in the nuclide). The even-mass-number abundances tend to be several times higher than those of odd-mass-number nuclides. In some mass ranges, the even-mass-number curve is zigzag whereas, with rare exceptions, the odd-mass-number curve is smooth. For this reason the odd-mass-number curve can be used to test data quality or to search for fraction-

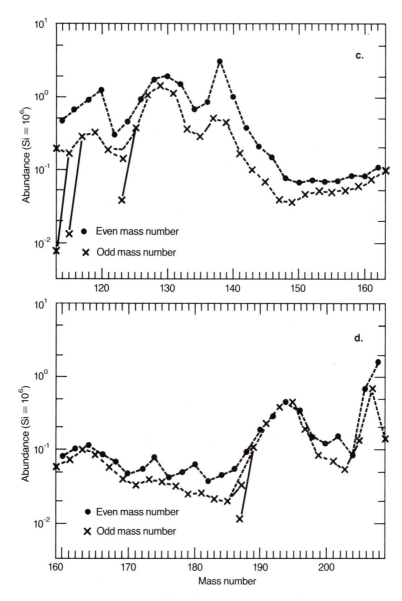

ations occurring during the formation of the CI chondrites. If an element has more than one stable odd-mass-number nuclide, these are connected by a solid line; such cases are valuable because they unambiguously establish the slope of the curve in these regions. In those few cases where two stable odd-mass-number nuclides have the same mass, both points are plotted, but the curve is drawn through the total abundance.

Table D-1

A new compilation of elemental concentration and abundance data for CI chondrites, and a comparison with abundances in the solar atmosphere

Atomic number	Element	Symbol	Solar abundance[°]	Abundance	CI chondrites		
					Concentration		References
1	Hydrogen	H	2.2×10^{10}	—[†]	20	mg/g	Wi69
2	Helium	He	2.3×10^9	—[†]	18	ng/g	Ma70
3	Lithium	Li	2.2×10^{-1}	6.05×10^1	1.57	μg/g	Ni74
4	Beryllium	Be	3.2×10^{-1}	8.01×10^{-1}	27	ng/g	An82
5	Boron	B	9.1×10^0	2.47×10^1	1.0	μg/g	An82, Cu80
6	Carbon	C	1.0×10^7	—[†]	32	mg/g	Wi69
7	Nitrogen	N	2.2×10^6	—[†]	1.5	mg/g	In74, Ke83, Ro82
8	Oxygen	O	1.9×10^7	—[†]	460	mg/g	Wi69
9	Fluorine	F	8.1×10^2	9.01×10^2	64	μg/g	Dr79, Go74
10	Neon	Ne	2.6×10^6	—[†]	300	pg/g	Ma70
11	Sodium	Na	4.7×10^4	5.59×10^4	4.8	mg/g	Ka81
12	Magnesium	Mg	9.6×10^5	1.07×10^6	97	mg/g	Ah69, Ka81, Wi69
13	Aluminum	Al	7.4×10^4	8.53×10^4	8.6	mg/g	Dr80, Ka81, Lo69, Wa74
14	Silicon	Si	$\equiv 1.0 \times 10^6$	$\equiv 1.00 \times 10^6$	105	mg/g	Ah69, Kn81, Wi69
15	Phosphorus	P	6.9×10^3	8.81×10^3	1.02	mg/g	Ah69
16	Sulfur	S	3.9×10^5	4.9×10^5	59	mg/g	Wi69
17	Chlorine	Cl	7.1×10^3	5.13×10^3	680	μg/g	Dr79
18	Argon	Ar	8.5×10^4	—[†]	1.3	ng/g	Ma70
19	Potassium	K	3.2×10^3	3.83×10^3	560	μg/g	Ah69, Ka81, Mi79, Ni74
20	Calcium	Ca	5.0×10^4	6.14×10^4	9.2	mg/g	Ah69, Ka81
21	Scandium	Sc	2.8×10^1	3.45×10^1	5.8	μg/g	Ka81, Sc64
22	Titanium	Ti	2.6×10^3	2.35×10^3	420	μg/g	Ah69, Dr80, Wa74, Wi69

			NRV				
23	V	Vanadium	3.0×10^2	2.94×10^2	56	μg/g	Ka81
24	Cr	Chromium	1.3×10^4	1.36×10^4	2.65	mg/g	Ka81
25	Mn	Manganese	6.0×10^3	9.25×10^3	1.90	mg/g	Ah69, Ka81, Pa78
26	Fe	Iron	8.5×10^5	8.72×10^5	182	mg/g	Ah69, Ka81
27	Co	Cobalt	2.0×10^3	2.31×10^3	508	μg/g	Ka81, Sc72
28	Ni	Nickel	4.3×10^4	4.88×10^4	10.7	mg/g	Ch76, Eb82, Ka81
29	Cu	Copper	2.6×10^2	5.09×10^2	121	μg/g	Kn81, Sc72, Wa74
30	Zn	Zinc	8.7×10^2	1.28×10^3	312	μg/g	Ch76, Eb82, Ia78, Ka81, Kr73
31	Ga	Gallium	1.4×10^1	3.76×10^1	9.8	μg/g	Ch76, Fo67a, Kn81
32	Ge	Germanium	7.1×10^1	1.22×10^2	33	μg/g	Ch76, Eb82, Fo67a, Ka81, Kr73, Ta78
33	As	Arsenic	8.9×10^0	6.57×10^0	1.84	μg/g	Ca72, Fo67b, Ka81
34	Se	Selenium	1.8×10^1	6.64×10^1	19.6	μg/g	Eb82, Ka81, Kr73, Ta78
35	Br	Bromine		1.21×10^1	3.6	μg/g	Go67, Ka81, Ta78
36	Kr	Krypton	4.7×10^1	——†	33	pg/g	Ma70
37	Rb	Rubidium	8.9×10^0	6.95×10^0	2.22	μg/g	Mi79
38	Sr	Strontium	1.8×10^1	2.41×10^1	7.9	μg/g	Ka70, Kn81, Mi79
39	Y	Yttrium	4.0×10^0	4.33×10^0	1.44	μg/g	Sc64
40	Zr	Zirconium	1.0×10^1	1.11×10^1	3.8	μg/g	Ga76, Kn81, Sh79, Wa74
41	Nb	Niobium	1.8×10^0	7.77×10^{-1}	270	ng/g	Gr72
42	Mo	Molybdenum	1.9×10^0	2.57×10^0	920	ng/g	Pa81
43	Tc	Technetium‡	——	——	——		——
44	Ru	Ruthenium	1.5×10^0	1.88×10^0	710	ng/g	Cr67, Ka81
45	Rh	Rhodium	5.6×10^{-1}	3.48×10^{-1}	134	ng/g	An82
46	Pd	Palladium	7.1×10^{-1}	1.41×10^0	560	ng/g	Eb82, Fo67b, Ka81
47	Ag	Silver	1.6×10^{-1}	5.16×10^{-1}	208	ng/g	Eb82, Kr73, Ta78
48	Cd	Cadmium	1.6×10^0	1.55×10^0	650	ng/g	Ch76, Eb82, Ka81, Kr73, Ta78
49	In	Indium	1.0×10^0	1.86×10^{-1}	80	ng/g	Ch76, Eb82, Fo67a, Ka81, Ta78

Table D-1 (*continued*)

Atomic number	Element	Symbol	Solar abundance[*]	CI chondrites Abundance	CI chondrites Concentration		CI chondrites References
50	Tin	Sn	2.0×10^{0}	3.88×10^{0}	1.72	$\mu g/g$	Eb82, Ha69, Kn81
51	Antimony	Sb	2.2×10^{-1}	3.36×10^{-1}	153	ng/g	Ca72, Eb82, Fo67b, Ka81, Kn81, Kr73
52	Tellurium	Te	NRV	5.03×10^{0}	2.4	$\mu g/g$	Eb82, Sm77
53	Iodine	I	NRV	1.05×10^{0}	500	ng/g	Dr79, Go67, Re66
54	Xenon	Xe	5.4×10^{0}	——†	32	pg/g	Ma70
55	Cesium	Cs	$<1.8 \times 10^{0}$	3.68×10^{-1}	183	ng/g	Eb82, Ke73, Mi79, Pa78, Sm64, Ta78
56	Barium	Ba	2.8×10^{0}	4.48×10^{0}	2.3	$\mu g/g$	Kn81, Na74, Re67
57	Lanthanum	La	3.0×10^{-1}	4.55×10^{-1}	236	ng/g	An82, Ev78
58	Cerium	Ce	7.9×10^{-1}	1.18×10^{0}	616	ng/g	An82, Ev78
59	Praseodymium	Pr	1.2×10^{-1}	1.76×10^{-1}	92.9	ng/g	An82, Ev78
60	Neodymium	Nd	3.8×10^{-1}	8.48×10^{-1}	457	ng/g	An82, Ev78
61	Promethium‡	Pm	——	——	——		——
62	Samarium	Sm	1.4×10^{-1}	2.65×10^{-1}	149	ng/g	An82, Ev78
63	Europium	Eu	1.1×10^{-1}	9.86×10^{-2}	56.0	ng/g	An82, Ev78
64	Gadolinium	Gd	3.0×10^{-1}	3.35×10^{-1}	197	ng/g	An82, Ev78
65	Terbium	Tb	NRV	5.98×10^{-2}	35.5	ng/g	An82, Ev78
66	Dysprosium	Dy	2.6×10^{-1}	4.03×10^{-1}	245	ng/g	An82, Ev78
67	Holmium	Ho	NRV	8.87×10^{-2}	54.7	ng/g	An82, Ev78
68	Erbium	Er	1.3×10^{-1}	2.56×10^{-1}	160	ng/g	An82, Ev78
69	Thulium	Tm	4.1×10^{-2}	3.91×10^{-2}	24.7	ng/g	An82, Ev78
70	Ytterbium	Yb	1.8×10^{-1}	2.46×10^{-1}	159	ng/g	An82, Ev78
71	Lutetium	Lu	1.3×10^{-1}	3.75×10^{-2}	24.5	ng/g	An82, Ev78
72	Hafnium	Hf	1.7×10^{-1}	1.80×10^{-1}	120	ng/g	Ga76, Kn81, Sh79, Wa74
73	Tantalum	Ta	NRV	2.51×10^{-2}	16	ng/g	Dr80
74	Tungsten	W	1.1×10^{0}	1.46×10^{-1}	100	ng/g	Wa74

76	Osmium	Os	1.1×10^{-1}	6.89×10^{-1}	490	ng/g	Eb82, Ka81, Ta78
77	Iridium	Ir	6.5×10^{-1}	6.40×10^{-1}	460	ng/g	Ch76, Eb82, Ka81, Kr73, Pa81, Ta78
78	Platinum	Pt	1.3×10^{0}	1.36×10^{0}	990	ng/g	Cr67, Eb82, Eh72, Kn81, Wa74
79	Gold	Au	1.3×10^{-1}	1.96×10^{-1}	144	ng/g	Ka81, Kr73, Ta78, Wa74
80	Mercury	Hg	$<2.8 \times 10^{0}$	5.20×10^{-1}	390	ng/g	An82, Re67
81	Thallium	Tl	1.8×10^{-1}	1.86×10^{-1}	142	ng/g	Eb82, Kr73, Re50, Ta78
82	Lead	Pb	1.9×10^{0}	3.10×10^{0}	2.4	µg/g	Ta76
83	Bismuth	Bi	$<1.8 \times 10^{0}$	1.41×10^{-1}	110	ng/g	Kr73, La70, Ta78
84–89	Unstable elements		—	—	—		
90	Thorium	Th	3.5×10^{-2}	4.34×10^{-2}	29	ng/g	Mo68, Ta76
91	Protactinium‡	Pa	$<8.9 \times 10^{-2}$	—	—		
92	Uranium	U	$<8.9 \times 10^{-2}$	9.22×10^{0}	8.2	ng/g	Kr73, Ta76, Ta78

* Solar-abundance data from J. E. Ross and L. H. Aller, *Science* 191:1223 (1976), as updated by L. H. Aller, to be published in *Spectroscopy and Astrophysical Plasmas*, eds. A. Dalgarno and D. Layzer, University of Cambridge, 1985. NRV = No reliable value; CI-chondrite abundances provide the best-available estimates of solar abundances for these elements.

† Volatile element for which no abundance was calculated. These elements are severely depleted in CI chondrites.

‡ Unstable element.

References:

Ah69   Ahrens, L., et al. 1969. *Earth Planet. Sci. Lett.* **6**:285.
An82   Anders, E., and M. Ebihara. 1982. *Geochim. Cosmochim. Acta* **46**:2363.
Ca72   Case, D., et al. 1972. *Geochim. Cosmochim. Acta* **36**:19–33.
Ch76   Chou, C.-L., et al. 1976. *Geochim. Cosmochim. Acta* **40**:85–94.
Cr67   Crocket, J. H., et al. 1967. *Geochim. Cosmochim. Acta* **31**:1615–1623.
Cu80   Curtis, D., et al. 1980. *Geochim. Cosmochim. Acta* **44**:1945–1953.
Dr79   Dreibus, G., et al. 1979. *Origin and Distribution of the Elements*, pp. 33–38. Pergamon.
Dr80   Dreibus, G., and H. Wänke. 1980. *A. Naturforsch.* **35a**:204–216.

Table D-1 (*continued*)

Eb82   Ebihara, M., et al. 1982. *Geochim. Cosmochim. Acta* **46**:1849–1862.
Eh72   Ehmann, W. D., and D. E. Gillum. 1972. *Chem. Geol.* **9**:1–11.
Ev78   Evensen, N. M., et al. 1978. *Geochim. Cosmochim. Acta* **42**:1199–1212.
Fo67a   Fouché, K. F., and A. A. Smales. 1967. *Chem. Geol.* **2**:5–33.
Fo67b   Fouché, K. F., and A. A. Smales. 1967. *Chem. Geol.* **2**:105–134.
Ga76   Ganapathy, R., et al. 1976. *Earth Planet. Sci. Lett.* **29**:302–308.
Go67   Goles, G. G., et al. 1967. *Geochim. Cosmochim. Acta* **31**:1771–1787.
Go74   Goldberg, R. H., et al. 1974. *Meteoritics* **9**:347–348.
Gr72   Graham, A. L., and B. Mason. 1972. *Geochim. Cosmochim. Acta* **36**:917–922.
Ha69   Hamaguchi, H., et al. 1969. *Geochim. Cosmochim. Acta* **33**:507–518.
In75   Injerd, W. G., and I. R. Kaplan. 1975. *Meteoritics* **9**:352–353.
Ka70   Kaushal, S. K., and G. W. Wetherill. 1970. *J. Geophys. Res.* **75**:463–468.
Ka81   Kallemeyn, G. W., and J. T. Wasson. 1981. *Geochim. Cosmochim. Acta* **45**:1217–1230.
Ke83   Kerridge, J. R. 1983. In preparation.
Kn81   Knab, H.-J. 1981. *Geochim. Cosmochim. Acta* **45**:1563–1572.
Kr73   Krähenbühl. U., et al. 1973. *Geochim. Cosmochim. Acta* **37**:1353–1370.
La70   Lal, J. C., et al. 1970. *Geochim. Cosmochim. Acta* **34**:89–103.
Lo69   Loveland, W., et al. 1969. *Geochim. Cosmochim. Acta* **33**:375–385.
Mi79   Mittlefehdt. D. W., and G. W. Wetherill. 1979. *Geochim. Cosmochim. Acta* **43**:201–206.
Mo67   Morgan, J. W., and J. F. Lovering. 1967. *Geochim. Comochim. Acta* **31**:1893–1909.

Table D-1 *(continued)*

Mo68 Morgan, J. W., and J. F. Lovering. 1968. *Talanta* **15**:1079 – 1095.

Na74 Nakamura, N. 1974. *Geochim. Cosmochim. Acta* **38**:757 – 775.

Ni74 Nichiporuk, W., and C. B. Moore. 1974. *Geochim. Cosmochim. Acta* **38**:1691 – 1701.

Pa78 Palme, H., et al. 1978. *Proc. Lunar Planet. Sci. Conf. 9th*, pp. 25 – 57.

Pa81 Palme, H., and W. Rammensee. 1981. *Earth Planet. Sci. Lett.* **55**:356 – 362.

Re60 Reed, G. W., et al. 1960. *Geochim. Cosmochim. Acta* **20**:122 – 140.

Re66 Reed, G. W., and R. D. Allen, Jr. 1966. *Geochim. Cosmochim. Acta* **30**:779 – 800.

Re67 Reed, G. W., and S. Jovanovic. 1967. *J. Geophys. Res.* **72**:2219 – 2228.

Ro82 Robert, F., and S. Epstein. 1982. *Geochim. Cosmochim. Acta* **46**:81 – 95.

Sc64 Schmitt, R. A., et al. 1964. *Geochim. Cosmochim. Acta* **28**:67 – 86.

Sc72 Schmitt, R. A., et al. 1972. *Meteoritics* **7**:131 – 213.

Sh79 Shima, M. 1969. *Geochim. Cosmochim. Acta* **43**:353 – 362.

Sm64 Smales, A. A., et al. 1964. *Geochim. Cosmochim. Acta* **28**:209 – 233.

Sm77 Smith, C. L., et al. 1977. *Geochim. Cosmochim. Acta* **41**:676 – 681.

Ta76 Tatsumoto, M., et al. 1976. *Geochim. Cosmochim. Acta* **40**:617 – 634.

Ta78 Takahashi, H., et al. 1978. *Geochim. Cosmochim. Acta* **42**:97 – 106.

Wa74 Wänke, H., et al. 1974. *Earth Planet. Sci. Lett.* **23**:1 – 7.

Wi69 Wiik, H. B. 1969. *Commun. Phys.-Math.* (Helsinki) **34**:135 – 145.

APPENDIX E

Radionuclides Used to Date Meteoritic Events

Table E-1 lists the long-lived and extinct radionuclides whose decay reactions are used to date meteoritic events, their stable daughter nuclides that exist in measurable abundances, their half-lives, and their decay constants.

Table E-1
Properties of radionuclides used to date meteoritic events

Parent	Measurable stable daughter(s)	Half-life $T_{1/2}$		Decay constant (yr^{-1})
Long-lived radionuclides:				
^{40}K	^{40}Ar, ^{40}Ca	1.25	Gyr	5.55×10^{-10}
^{87}Rb	^{87}Sr	48.8	Gyr	1.42×10^{-11}
^{147}Sm	^{143}Nd	106.	Gyr	6.54×10^{-12}
^{187}Re	^{187}Os	~50.	Gyr	$\sim 1.4 \times 10^{-11}$
^{232}Th	^{208}Pb, ^4He	14.0	Gyr	4.95×10^{-11}
^{235}U	^{207}Pb, ^4He	0.704	Gyr	9.85×10^{-10}
^{238}U	^{206}Pb, ^4He	4.47	Gyr	1.55×10^{-10}
Extinct radionuclides:				
^{26}Al	^{26}Mg	0.72	Myr	9.6×10^{-7}
^{107}Pd	^{107}Ag	6.5	Myr	1.07×10^{-7}
^{129}I	^{129}Xe	16.	Myr	4.3×10^{-8}
^{244}Pu	$^{131-136}Xe°$	82.	Myr	8.5×10^{-9}

Source: Adapted from T. Kirsten, in *The Origin of the Solar System*, ed. S. F. Dermott, p. 267. Wiley, 1978.
° Many other fission products also are produced.

APPENDIX F

Calculation of Equilibrium Relationships in the Solar Nebula

Insight into the fractionations observed in chondritic meteorites can result from the comparison of elemental abundances with those expected if stable solid phases were gained or lost from a cooling solar nebula. The solid phases that form by equilibrium processes in the solar nebula can be determined by following the principles of chemical thermodynamics. Consider the simple chemical reaction

$$a\,A + b\,B = c\,C + d\,D \tag{F-1}$$

in which a moles of substance A react with b moles of substance B to produce c moles of substance C and d moles of substance D. At any temperature we can write an **equilibrium constant** K for this reaction:

$$K = \frac{C^c D^d}{A^a B^b} \tag{F-2}$$

where (if the substances behave ideally) A, B, C, and D represent pressures for gaseous substances or concentrations for solid or liquid[1] substances. If, as is generally true, the substances behave nonideally, then a corrected pressure called **fugacity** is used for gases, and a corrected concentration called **activity** is used for solids and liquids.

Thermochemical data are commonly tabulated in terms of various functions. One of these, the Gibbs free energy G, is related to the equilibrium constant K as

$$\Delta G = -RT \ln K \tag{F-3}$$

where T is the absolute temperature, R is the gas constant, and the Δ in front of the G means "the change in" G during the reaction as written. Gibbs-free-energy data are commonly tabulated as the free energies of formation ΔG_f from the elements in their standard states (their states at a temperature

---

[1] No liquids form as stable phases in a low mass-solar nebula having $p_{H_2} < 10^{-2}$ atm.

of T and a pressure of 1 atm). For the reaction in Equation F-1,

$$\Delta G = c \, \Delta G_{f,C} + d \, \Delta G_{f,D} - a \, \Delta G_{f,A} - b \, \Delta G_{f,B} \qquad \text{(F-4)}$$

The first step in calculating a condensation temperature for an element is to choose the set of solar-abundance data that is to be used—for example, those in Appendix D (CI-chondrite values where available, solar-atmosphere values for the most-volatile elements). Next, one must choose a pressure for the nebula. Because most of the mass is hydrogen, and because the most stable form of hydrogen at relevant pressures and temperatures is H_2, the total nebula pressure can be set equal to p_{H_2}, the pressure of H_2. For a minimum-mass nebula having a total mass of 0.01 $M_\odot$, the amounts of material in the inner planets imply H_2 pressures at the nebula midplane ranging from about 10^{-4} atm at 0.5 AU to 10^{-6} or 10^{-7} atm at 4 AU (see Figure VIII-1).

The partial pressures of the major gaseous species for each element should now be determined. For example, examination of the relative stabilities of many gaseous species including O, O_2, OH, SiO, and CO_2 show that the chief species of oxygen are $H_2O(g)$ and $CO(g)$.[2] The relative abundance of these two species can be determined by solving a series of simultaneous equations. For example, complete studies show that, under most nebular conditions, the two dominant gaseous species of carbon are CH_4 and CO. We can then write

$$CO(g) + 3 \, H_2(g) = CH_4(g) + H_2O(g) \qquad \text{(F-5)}$$

The calculations would typically be carried out for a series of different H_2 pressures. Because we have three unknowns p_{CO}, p_{CH_4}, and p_{H_2O}, we require three equations. One is the equilibrium relationship:

$$K = \frac{p_{CH_4} p_{H_2O}}{p_{CO} p_{H_2O}^3} \qquad \text{(F-6)}$$

where $p_{H_2O}^3$ means $(p_{H_2O})^3$. The other two equations are based on conservation of mass. If carbon is either CH_4 or CO, then

---

[2] The letter in parentheses gives the state: g = gas, l = liquid, s = solid. In some cases, the crystalline state of the solid must be specified—for example, quartz, tridymite, or cristobalite for SiO_2.

$$p_{CO} + p_{CH_4} = 2\,\frac{h_C}{h_H}\,p_{H_2} \tag{F-7}$$

where h_C and h_H are the solar abundances[3] of carbon and hydrogen, and the 2 results from the fact that there are two hydrogen atoms per molecule. Similarly,

$$p_{H_2O} + p_{CO} = 2\,\frac{h_O}{h_H}\,p_{H_2} \tag{F-8}$$

The general relationship upon which relationships F-7 and F-8 are based is

$$a p_{X_aQ} + b p_{X_bR} + c p_{X_cT} + \cdots = 2\,\frac{h_X}{h_H}\,p_{H_2} \tag{F-9}$$

where a is the number of X atoms in the gaseous compound X_aQ, Q includes all the remaining atoms in this species, and so on.

Condensation begins when the nebula cools to the point that the pressure of a gaseous species of element X just exceeds the pressure in equilibrium with the most-stable solid form of X. Consider the following simple example based on iron. Relationships similar to those just described show that the dominant gaseous species is $Fe(g)$, and that the most-stable solid species is $Fe(s)$.

The condensation reaction is then

$$Fe(s) = Fe(g) \tag{F-10}$$

For convenience, the reaction is written backward in order to have p_{Fe} in the numerator of the equilibrium expression:

$$K = \frac{p_{Fe}}{c_{Fe}} \tag{F-11}$$

where c_{Fe} is the concentration of Fe in the solid. Concentrations in solids or

---

[3] Because a is used for activity, we use h as the symbol for abundance, from the equivalent German word, *Häufigkeit*.

liquids are generally expressed as mole fractions, the number of moles of the element divided by the total numbers of moles of all elements. In fact, iron does form an alloy with nickel, but nickel in bulk solar matter is 19 times less abundant than iron, and the mole fraction of Fe in the condensing solid is near 0.9 under most sets of plausible nebula conditions.

Condensation temperatures are commonly tabulated as the temperature at which the abundance of an element in its gaseous forms (see Equation F-9) falls to 0.5 times the initial value. Such 50-percent condensation temperatures for iron are 1336 K and 1185 K at nebular pressures of 10^{-4} and 10^{-6} atm, respectively.

The condensation of a trace element in Fe–Ni is handled analogously. For example, Au condenses by the reaction

$$Au(s) = Au(g) \tag{F-12}$$

with

$$K = \frac{p_{Au}}{c_{Au}} \tag{F-13}$$

Available thermochemical data indicate that Au(s) condenses in solid solution in Fe–Ni(s). We again use the mole fraction as the measure of c_{Au}. Because trial calculations show that gold condenses after iron and nickel, the Au mole fraction at the 50-percent condensation temperature is 0.5 $h_{Au}/(h_{Fe} + h_{Ni})$, where h again represents solar abundance.

Here we run into a complication. Metallurgical studies show that gold does not form an ideal solution in Fe–Ni, but that the gold activity a_{Au} is about 10 times larger than c_{Au}, the correction factor (called the **activity coefficient**) increasing with decreasing temperature. After making these corrections, the 50-percent condensation temperatures of gold are found to be 1225 K and 1074 K at 10^{-4} and 10^{-6} atm, respectively.

In many nebular reactions, the element is in a different oxidation state in the main gaseous species than in the condensed form. For example,

$$Ge(s) + H_2O(g) = GeO(g) + H_2(g) \tag{F-14}$$

and

$$K = \frac{p_{GeO} p_{H_2}}{c_{Ge} p_{H_2O}} \tag{F-15}$$

The equilibrium expression is similar to those described previously, but it has two more terms. However, because $H_2O(g)$ is a far-more-abundant species than $GeO(g)$ or any other species undergoing condensation, the ratio p_{H_2}/p_{H_2O} remains essentially constant during the reaction. In contrast to the case described for gold, the activity coefficient for germanium dissolved in Fe – Ni is less than unity. Estimated values are in the range of 10^{-2} to 10^{-3}, and they decrease with decreasing temperature.

A final example involves reaction with previously condensed phases. The most abundant gaseous species of sulfur is H_2S, and the condensation reaction is

$$FeS(s) + H_2(g) = H_2S(g) + Fe(s) \qquad \text{(F-16)}$$

with the K expression

$$K = \frac{c_{Fe}p_{H_2S}}{c_{FeS}p_{H_2}} \qquad \text{(F-17)}$$

The value of c_{FeS} is unity, because FeS is not appreciably diluted by other species. Because c_{Fe} is always ~ 0.9, the 50-percent condensation temperature is that at which

$$K = 0.9\,\frac{p_{H_2S}}{p_{H_2}} \cong 0.9\,\frac{h_S}{h_H} \qquad \text{(F-18)}$$

(note that I have cancelled 0.5 in the numerator (50 percent condensation of sulfur) and in the denominator (two H per H_2). Because the nebular pressure does not appear in this expression, the equilibrium condensation temperature (648 K) is independent of nebular pressure.

Present in chondrites of all classes are phases that would not have survived if equilibrium had been maintained down to moderately low temperatures. The refractory inclusions in the CV chondrites are a good example; their calcium, aluminum and probably also titanium should entirely enter pyroxene and plagioclase as soon as these phases formed if equilibrium were maintained. That the inclusions did survive almost certainly indicates either (1) that the combination of their large size and limited diffusion distances prevented equilibration with surrounding minerals, or (2) that the refractories and the silicates formed separate nebular grains that could not exchange material because of the very-low pressures of the gaseous species of the constituent elements.

Returning to the case of the condensation of sulfur as FeS and noting that the nebular abundance of sulfur is about 0.6 times that of iron (see Appendix D), we see that complete condensation of sulfur by Reaction F-16 requires the conversion of more than one-half of the Fe–Ni to FeS. In most classes of chondrites, the abundance of sulfur is much lower than CI values, and it seems likely that there was incomplete reaction of the H_2S with Fe–Ni grains as a result of limited diffusion of iron from the Fe–Ni to the surface of the FeS "corrosion" layer on the outside of the grains. When such kinetic effects were present, true 50-percent condensation temperatures may have been significantly lower than equilibrium values. It seems probable that at some nebula locations the temperature was less than 600 K when 50 percent of the sulfur had condensed, and that an appreciable fraction of the sulfur condensed by reactions other than the indicated "equilibrium" reaction with Fe–Ni.

Calculations of nebular equilibria are still in a relatively immature stage of development. The appropriate thermochemical data are not always available. Even when they are available, the researchers doing the calculations sometimes overlook important phases. It is therefore useful to note the effects of failing to include relevant phases in a set of calculations. Neglect of a significant gaseous phase of the element results in a condensation temperature that is erroneously high; neglect of a significant solid phase results in a condensation temperature that is erroneously low. For those elements that form solid solutions, assumptions of ideal behavior (activity coefficient of 1) yield too low a temperature if the activity coefficient is less than unity and too high a temperature if the correct value is greater than unity.

A major difficulty for many elements is locating appropriate thermochemical data. If the data are not in standard compilations such as those listed below, it will usually be necessary to search *Chemical Abstracts* and, if one still cannot find sources of the needed data, to make educated guesses using techniques such as those given in Chapter III of Kubachewski and Alcock (1979).

Compilations of Thermochemical Data

Hultgren, R., et al. 1973. *Selected Values of the Thermodynamic Properties of the Elements.* Amer. Soc. Metals.

Hultgren, R., et al. 1973. *Selected Properties of the Thermodynamic Properties of Binary Alloys.* Amer. Soc. Metals.

JANAF. 1971. (*Thermochemical Tables,* ed. D. R. Stull and H. Prophet),

Nat. Stand. Ref. Data Ser. **37.** National Bureau of Standards (U.S.). Also later supplements in *J. Phys. Chem. Ref. Data* (e.g., 7:793–940, 1978).

Kubachewski, O., and C. B. Alcock. 1979. *Metallurgical Thermochemistry.* Pergamon.

Robie, R. A., B. S. Hemingway, and J. R. Fisher. 1978. Thermodynamic properties of minerals and related substances at 298.15 K and 1 bar (10^5 pascals) and at higher temperatures. *Geol. Surv. Bull.* 1425.

APPENDIX G

Nebular 50-Percent Condensation Temperatures

Table G-1 lists 50-percent condensation temperatures (see Appendix F) calculated for a cooling mixture of the elements in solar abundances at pressures of 10^{-4}, 10^{-5}, and 10^{-6} atm. Complete gaseous equilibrium and limited gas–solid equilibrium are assumed. Only elements having 50-percent condensation temperatures of 400 K or greater are listed in the table.

Table G-1
Nebular 50-percent condensation temperatures

Element	50-percent condensation temperature (K)			References
	10^{-4} atm	10^{-5} atm	10^{-6} atm	
Li	1225	——	1091	Wa77
Be	——	——	——	——
B	——	——	——	——
F	736	729	721	Fe80; see also Wa77
Na	970	911	861	Fe80; see also Wa77, Gr74a
Mg	1340	1268	1203	Wa78b; see also Gr74a
Al	1650	1591	1531	La84; see also Gr74a
Si	1311	1251	1193	La84; see also Gr74a
P	1151	1110	1070	Fe80; see also Se78, Wa77
S	648	648	648	Wa77; see also Se78
Cl	863	817	775	Fe80
K	1000	941	890	Fe80; see also Wa77, Gr74a
Ca	1518	1447	1382	Wa78b; see also Gr74a
Sc	1644	——	——	Gr74a
Ti	1549	1482	1420	Wa78b
V	~1450	~1380	~1310	Gr74a
Cr	1277	——	1137	Wa78a; see also Gr74b, Fe84
Mn	1190	——	1078	Wa77
Fe	1336	——	1183	Wa81; see also Se78, Pa76, Fe84
Co	1351	——	1197	Wa81; see also Pa76, Fe84
Ni	1354	——	1202	Wa81; see also Se78, Pa76, Fe84
Cu	1037	——	910	Wa79

Element	50-percent condensation temperature (K)			References
	10^{-4} atm	10^{-5} atm	10^{-6} atm	
Zn	660	——	605	Wa77
Ga	918	——	738	Wa79; see also Se78
Ge	825	——	702	Wa79; see also Se78
As	1157	——	1012	Wa79
Se	684	684	684	Wa77
Br	~690	~690	~690	Ka85; see also Fe80
Rb	~1080	——	——	Gr74a
Sr	——	——	——	——
Y	1592	——	——	Bo81
Zr	~1780	——	——	Gr74a
Nb	~1550	——	——	Gr74a
Mo	1608	——	1456	Wa81; see also Pa76, Fe84
Ru	1573	——	1437	Wa81; see also Pa76, Fe84
Rh	1391	——	1255	Wa81; see also Pa76, Fe84
Pd	1334	——	1175	Wa78a, Fe84
Ag	952	——	843	Wa77
Cd	——	430	——	An76
In	——	456	——	Gr74a
Sn	720	——	625	Wa77
Sb	912	——	767	Wa79
Te	680	——	600	Wa77
I	——	——	——	——
Cs	——	——	——	——
Ba	——	——	——	——
La	1520	——	——	Bo80
Ce	1500	——	——	Bo80
Pr	1532	——	——	Bo80
Nd	1510	——	——	Bo80
Sm	1515	——	——	Bo80
Eu	1450	——	——	Bo80
Gd	1545	——	——	Bo80
Tb	1560	——	——	Bo80
Dy	1571	——	——	Bo80
Ho	1568	——	——	Bo80
Er	1590	——	——	Bo80
Tm	1545	——	——	Bo80
Yb	1455	——	——	Bo80

Table G-1 *(continued)*

Element	50-percent condensation temperature (K)			References
	10^{-4} atm	10^{-5} atm	10^{-6} atm	
Lu	1597	——	——	Bo80
Hf	1652	——	——	Gr74a
Ta	~1550	——	——	Gr74a
W	1802	——	1631	Wa81; see also Gr74a, Pa76, Fe84
Re	1819	——	1669	Wa81; see also Gr74a, Pa76, Fe84
Os	1814	——	1666	Wa81; see also Gr74a, Pa76, Fe84
Ir	1610	——	1464	Wa81; see also Pa76, Fe84
Pt	1411	——	1278	Wa81; see also Pa76, Fe84
Au	1225	——	1074	Wa79
Hg	——	——	——	——
Tl	——	428	——	Gr74a
Pb	——	496	——	Gr74a
Bi	——	451	——	Gr74a
Th	1545	——	——	Bo81; see also Gr74a
U	1420	——	——	Bo81; see also Gr74a
Pu	1520	——	——	Bo81; see also Gr74a
Cm	1530	——	——	Bo81

References:

An76   Anders, E., et al. 1976. *Geochim. Cosmochim. Acta* **40**:1131.
Bo80   Boynton, W. V. 1980. Personal communication.
Bo81   Boynton, W. V. 1981. Personal communication.
Fe80   Fegley, B. 1980. Personal communication.
Fe84   Fegley, B., and H. Palme. 1984. *Earth Planet. Sci. Lett.* (submitted).
Gr74a   Grossman, L., and J. W. Larimer. 1974. *Rev. Geophys. Space Phys.* **12**:71.
Gr74b   Grossman, L., and E. Olsen. 1974. *Geochim. Cosmochim. Acta* **38**:173.
Ka85   Kallemeyn, G. W. and J. T. Wasson. 1985. *Rev. Geophys. Space Phys.* (submitted).
La84   Larimer, J. W. 1984. Personal communication.
Pa76   Palme, H., and F. Wlotzka. 1976. *Earth Planet. Sci. Lett.* **33**:45.
Se78   Sears, D. 1978. *Earth Planet. Sci. Lett.* **41**:128.
Wa77   Wai, C. M., and J. T. Wasson. 1977. *Earth Planet. Sci. Lett.* **36**:1.
Wa78a   Wai, C. M., et al. 1978. *Lunar Planet. Sci.* **9**:1193.
Wa78b   Wasson, J. T. 1978. In *Protostars and Planets*, ed. T. Gehrels, pp. 478–501. University of Arizona Press.
Wa79   Wai, C. M., and J. T. Wasson. 1979. *Nature* **282**:790.
Wa81   Wai, C. M. 1981. Personal communication.

APPENDIX H

Rudiments of Celestial Mechanics

The motion of a body in orbit about another body describes an ellipse. The properties of an ellipse are summarized in Figure H-1a. The long axis is called the major axis, the short axis the minor axis. The most-commonly-used property of an orbital ellipse is one-half of the major axis, called the **semimajor axis** and generally represented by the symbol a. One-half of the minor axis is called the **semiminor axis** and represented by b.

As also shown in Figure H-1a, an ellipse has two foci. The sum of the distance from any point on the perimeter to each of the foci is $2a$; thus the distance from either focus to the outer end of a semiminor axis is a. From this fact and the relation between the sides of a right triangle, one can readily show that the distance f from the center to either focus is given by the relationship $f = (a^2 - b^2)^{1/2}$.

The eccentricity e is a common measure of the shape of the ellipse; it is defined by $e = f/a$ and can have any value between 0 and 1. If $e = 0$, the ellipse is a circle; if $e = 1$, the ellipse is a line segment.

The typical orbit relevant to this text is described by a smaller body in orbit about a much larger body — for example, a planet or meteoroid in orbit about the Sun. In such a case, the larger body occupies one focus of the elliptical orbit. An elliptical heliocentric orbit is shown in Figure H-1b. Two other quantities are defined to facilitate the discussion of such an orbit: the greatest distance from the Sun is called the **aphelion** Q, and the least distance from the Sun is called the **perihelion** q. Note that these distances combine to yield the major axis, so that $Q + q = 2a$. For the general case of an orbit about any massive object, these quantities are given the generic names apoapsis (still represented by Q) and periapsis (q). For a satellite of the Earth, the relevant names are apogee (Q) and perigee (q).

A body in heliocentric orbit moves fastest near perihelion, slowest near aphelion. This observation was first put into quantitative terms by Kepler, whose second law states that "a line between the planet and the Sun sweeps out equal areas in equal amounts of time." Thus, if the aphelion distance is 3 times greater than the perihelion distance, an object's orbital velocity will be 3 times greater at perihelion than at aphelion.

For a small object in a circular orbit about a large body, the orbital velocity v can be derived by equating the gravitational force $F_g = GMm/a^2$ with the centrifugal force $F_c = mv^2/a$, where M is the mass of the large body, m is the mass of the small body, and G is the universal gravitational constant. This equation yields the result

$$v = \left(\frac{GM}{a} \right)^{1/2} \qquad \text{(H-1)}$$

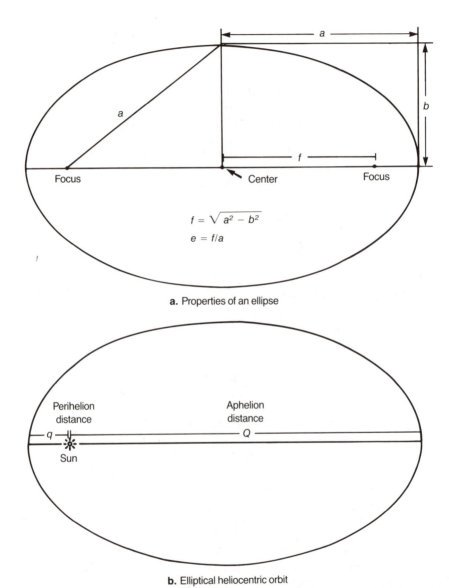

a. Properties of an ellipse

$$f = \sqrt{a^2 - b^2}$$
$$e = f/a$$

b. Elliptical heliocentric orbit

Figure H-1. (*a*) Sketch of an ellipse, showing the symbols representing the main parameters: semimajor axis *a*, semiminor axis *b*, two foci each a distance *f* from the center, and the eccentricity *e*. (*b*) In an elliptical heliocentric orbit, the Sun occupies one focus. The greatest distance from the Sun is called the aphelion *Q*, the shortest distance from the Sun the perihelion *q*. Together *Q* and *q* comprise the major axis of the ellipse.

Equation H-1 also is valid for the *mean* orbital velocity of an object in an elliptical orbit. The velocity at any point in an elliptical orbit is given by

$$v = \left[GM\left(\frac{1}{r} - \frac{1}{2a}\right) \right]^{1/2} \tag{H-2}$$

where r is the distance separating the small and large bodies.

The period p of a body in an elliptical orbit is given by $p = 2\pi a/v$, where v is the mean velocity. Substituting the value of v given by Equation H-1, we obtain

$$p = \frac{2\pi a^{1.5}}{G^{0.5}M^{0.5}} = Ka^{1.5} \tag{H-3}$$

In the final part of Equation H-3, the constant terms are combined to give a single constant K. For heliocentric orbits, K is unity if p is expressed in yr and a in AU. If the mass of the smaller object is not negligible relative to that of the more massive body, then one must substitute $(M + m)$ for M in Equation H-3. The relationship described by Equation H-3 is Kepler's third law. (His first law states that planetary motions describe elliptical orbits.)

The orbits of the planets are nearly coplanar with that of the Earth, but the typical angles between the orbital planes of asteroids and that of the Earth are about 10°, and the angles between the planes of long-period comets and that of the Earth can have any value between 0° and 180°.

This angle between the two orbital planes is called the **inclination** i. It is illustrated in Figure H-2. The most common reference plane is the plane of the Earth's orbit (the **ecliptic plane**), but for some purposes the mean plane of the solar system (essentially the same as the plane of Jupiter's orbit) is more useful.

Although the definition of inclination is commonly illustrated as the angle between planes as shown on a diagram such as Figure H-2, this representation is ambiguous because either of two angles (that sum to 180°) could be chosen. A more-exact definition of the inclination is the angle between the north polar vectors of the orbits, where the north polar vector is the perpendicular direction in which the thumb of the right hand points when the fingers point in the direction of an object's motion in its orbit.

In order for a meteoroid (or asteroid or comet) to enter the Earth's atmosphere, the meteoroid's orbit must intersect the orbit of the Earth. If the meteoroid's orbit is inclined to that of the Earth, the intersection can occur only along the **line of nodes** defined by the intersection of the two orbital planes (see Figure H-2). If the meteoroid is in a typical orbit having a moderate eccentricity, there normally can be only one point at which the

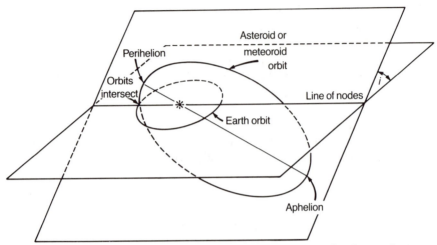

Figure H-2. In order for an object in heliocentric orbit to strike the Earth, its orbit must intersect that of the Earth. The angle between the planes of the two orbits is the inclination *i*. Intersection of the orbits can occur only along the line of nodes formed by the intersection of the orbital planes. Because of perturbations of the orbital elements by the planets, any particular Earth-crossing orbit is Earth-intersecting during only a small fraction of the time.

two orbits can intersect (the trivial exception occurs if the major axis of the meteoroid's orbit is perpendicular to the line of nodes). In most orbital passages, the meteoroid and the Earth do not reach the intersection point at the same time.

As a result primarily of gravitational perturbations by Jupiter, the orientation of a meteoroid's orbit undergoes significant changes on a time scale of 10 to 100 kyr. Although most objects in **Earth-crossing orbits** (those with $Q > 1$ AU and $q < 1$ AU) do not actually intersect the Earth's orbit, these changes in orientation cause the orbits to be intersecting a small fraction of the time. When one includes the effects of these long-distance perturbations, one finds that an Earth-crossing meteoroid (or larger object) about once each megayear makes a close approach to the Earth that causes a major change in the orbital parameters but leaves the object in an Earth-crossing orbit, because the object's new orbit must include the point where the close approach to the Earth occurred.

If an object is in an Earth-crossing orbit confined to the inner solar system ($Q < 4.5$ AU), it will not be able to come close enough to Jupiter to receive a major impulse. Numerical simulations show that the object will be removed from such an orbit after about 10 Myr. The most probable fates that await it are collision with the Earth, collision with Venus, or an orbital perturbation by one of these planets that makes it Jupiter-crossing and allows Jupiter to eject it from the solar system within about 100 kyr.

APPENDIX I

How to Recognize Meteorites

Conversations with both professional and lay persons reveal that they and I share a common fascination; when we take walks, we are continuously (though often subconsciously) visually examining the rocks along the trail for possible meteorites. I have never found a meteorite in this fashion, and less than 1 in 1000 persons who indulge in this habit will ever find a meteorite. It's fun nonetheless, and it is therefore useful to know what clues indicate that a strange rock might be a meteorite.

As discussed in some detail in Chapter II, most meteorites that are seen to fall are chondrites, the remainder being stony-irons and differentiated silicate-rich meteorites ($\sim$ 9 percent) and irons ($\sim$ 6 percent). About 85 percent of all meteorites contain at least 20 mg/g (2 wt%) metallic Ni–Fe and, because these metal-bearing meteorites are more resistant to weathering, the proportion of meteorite finds that are metal-bearing is at least 95 percent. Because metallic Fe–Ni is extremely rare among terrestrial rocks, the presence of such metal in a "rock" is the best indication that the rock is a meteorite.

A key question then is how can one most easily find out whether a rock contains metallic iron. Those with modern analytical equipment available will have no trouble showing that the metal is Fe–Ni; the most straightforward technique is to analyze suspicious minerals with the electron microprobe, a device that measures X rays emitted by the sample following its bombardment by electrons.

Those who do not have analytical facilities available can use a series of clues to tell them whether metallic Fe–Ni is present. Because of the high density of metal ($\sim$ 8 g $\cdot$ cm^{-3}), meteorites containing metal, with rare exceptions, have densities of at least 3.3 g $\cdot$ cm^{-3} (3.3 times the density of water), appreciably denser than other local rocks. The rare terrestrial rocks having densities greater than or equal to 3.3 g $\cdot$ cm^{-3} are generally iron ores such as hematite or magnetite. Density can be measured by weighing an object to determine the mass and dividing this by the volume, which can be determined by various techniques; the most accurate method is weighing the object again while suspended in a liquid such as water. The volume in cubic centimeters is numerically equal to the reduction in mass in grams that results from suspension in water.

Meteoritic metal responds to a magnet. If much metal is present, the magnet will cling to the rock; minor amounts of metal can be detected by hanging the magnet on a string and testing to see if the rock has enough attraction to cause the magnet to swing away from the vertical. Of course, human-produced metallic iron attracts a magnet equally strongly, and some iron ores such as magnetite also show an appreciable attraction.

Metallic iron in a typical chondrite will show up as shiny millimeter-sized spots on a flat surface produced by grinding. The combination of shiny metallic spots, a weak magnetic attraction, and a density of at least 3.3 $g \cdot cm^{-3}$ is strong evidence that a rock is a meteorite.

Human-produced iron is the material most likely to masquerade as an iron meteorite. In many cases, artificial iron has shapes (such as straight edges or 90° corners) distinctly different from those of meteorites (see Chapter II). Nickel, which is present at levels of at least 40 mg/g (4 wt%) in every iron meteorite, is present in human-made iron only as an uncommon additive — in stainless steel, for example. Stainless steel generally contains more chromium than nickel, however, whereas chromium in iron meteorites virtually never is present at levels as great as 1 mg/g. Artificial iron commonly contains other alloying elements such as vanadium, manganese, silicon, or carbon that are present at very low levels in meteoritic iron.

An iron meteorite can almost always be recognized by the texture revealed by etching a polished surface. In particular, most irons show the Widmanstätten pattern illustrated in Figures II-6, II-7, and II-8. This pattern is revealed by etching a finely polished surface with nital, a 2-percent solution of concentrated nitric acid in ethanol. Care is needed because this acid interacts strongly with human skin!

The external morphology of a meteorite is commonly characteristic. The wavelike surface seen on the Cabin Creek iron (Figure I-3) is found on many (but not all) iron and in a more subdued "thumbprinted" form on many stones. Some morphologies are virtually never found. For example, it appears (based on samples sent to me for identification) that many persons believe meteorites to have frothy, porous, or slaggy textures; in fact, meteorites are invariably compact, with no visually recognizable porosity.

Almost every freshly fallen meteorite has a dark fusion crust covering the surface. The blackness of the crust contrasts with the light gray of the interior of many stones. In some plagioclase-rich meteorites, the fusion crust is quite shiny (see Figure II-5). Unfortunately, the crust becomes the reddish color of iron rust within a few years; thus, in most cases, the color of meteorite finds is rather similar to those of other local rocks. Partly because of this, new stony meteorite finds are rarely recognized in rocky areas. It is not a coincidence that most meteorite finds in the U.S. are recovered from the fine, rock-free soil of the western plains.

If you have an object having properties indicating that it may be a meteorite, the best way to obtain confirmation is to send a small, walnut-size fragment to a major meteorite collection. You can send it to me at UCLA, Los Angeles, CA 90024, for free examination. The same service is offered in the U.S. by the American Museum of Natural History, New York, NY 10024; Arizona State University, Tempe, AZ 85281; the Field Museum of Natural History, Chicago, Il 60605; the Smithsonian Institution, Washington, DC 20560; and the University of New Mexico, Albuquerque, NM 87131. Other important world locations offering this service are the Geological Survey of Canada, Ottawa K1A 0E8 Ontario, Canada; the British Museum (Natural History), London SW7 5BD, England; and Muséum Histoire Naturelle, Paris 75005, France; the Naturhistorisches Museum, Vienna A-1014, Austria; the Max Planck Institut für Chemie, Mainz D-6500, Federal Republic of Germany; the Western Australian Museum, Perth, WA 6000, Australia; and the national scientific or natural-history museums of many other countries.

One of the best ways to learn to recognize meteorites is to visit meteorite exhibits at museums. Note in particular the morphologies and colors of the finds, especially the stony-meteorite finds (irons are much easier to recognize).

If you should ever have the grand fortune to see a meteorite streak through the sky, the chief piece of information you should note is the exact direction in which it disappeared. Point to and fix in your mind a distant landmark that is in line with the end of the fireball's path, and then note exactly where you are standing.

Index

Page numbers in **boldface** refer to a definition.